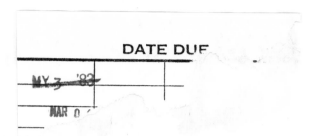

Factor Analysis in Chemistry

Factor Analysis in Chemistry

EDMUND R. MALINOWSKI

Professor of Chemistry
Stevens Institute of Technology

DARRYL G. HOWERY

Professor of Chemistry
City University of New York, Brooklyn College

A Wiley-Interscience Publication

John Wiley & Sons

New York Chichester Brisbane Toronto

Library of Congress Cataloging in Publication Data

Malinowski, Edmund R
    Factor analysis in chemistry.

    "A Wiley-Interscience publication."
    Bibliography: p.
    Includes index.
    1. Chemistry—Mathematics. 2. Factor analysis.
I. Howery, D. G., joint author. II. Title.

QD39.3.M3M34   542   79-27081
ISBN 0-471-05881-5

Printed in the United States of America

10 9 8 7 6 5 4 3 2 1

To Professor Luigi Z. Pollara,
who planted the seed.

Preface

The development of new methods for interpreting large data sets ranks as one of the major advances in chemistry during the past decade. Major weapons in this still evolving "chemometric revolution" are high-speed digital computers and a variety of mathematical–statistical methodologies.

Factor analysis, a mathematical technique for studying matrices of data, is one of the most powerful methods in the arsenal of chemometrics. The method has been used more than 50 years in the behavioral sciences, and has been applied successfully to chemical problems during the past 15 years. In this book we cover the theory, practice, and application of factor analysis in chemistry. Whereas previous publications have dealt solely with classical, "abstract" factor analysis, we shall stress the target-transformation version of factor analysis. The model-testing capabilities of "target" factor analysis enable the chemist to identify physically significant factors that influence data.

Although chemical applications are emphasized, the book should be of interest to physical scientists in general. In addition, the target-transformation approach should prove valuable to behavioral scientists.

Our main objectives are to explain factor analysis and thereby to facilitate greater use of this technique in chemical research. The qualitative material can be followed by persons having a background in physical chemistry, and the theoretical sections can be assimilated if the chemist has some familiarity with matrix algebra. We have four specific objectives:

1. To explain the essentials of factor analysis in qualitative terms.
2. To develop the mathematical formulation of factor analysis in a rigorous fashion.
3. To describe how to carry out a factor analysis.
4. To survey the applications of factor analysis in chemistry.

Readers can acquire a qualitative overview of factor analysis in Chapters 1 and 2. A detailed study of the mathematical formulation presented and illustrated in Chapters 3, 4, and 5 is necessary for a thorough understanding of factor analysis. Users of factor analysis will find practical instructions for carrying out an analysis in Chapter 6. Chapters 7 through 10 constitute a review of the many applications of factor analysis to chemical problems.

For the record, Chapters 1, 2, 6, and 9 are primarily the work of D.G.H.; Chapters 3, 4, 5, 7, and 8 are due almost entirely to E.R.M.; and Chapter 10 is a joint effort. We owe special thanks to Dr. Paul H. Weiner for useful discussions and for contributions to Chapters 8, 9, and 10. We also thank Matthew McCue for his excellent technical suggestions, comments, and corrections. For the tedious task of typing the many drafts of the manuscript, we are indebted to Louise Sasso, Joanne Higgins, and our wives, Helen M. and Yong H.

EDMUND R. MALINOWSKI
DARRYL G. HOWERY

Hoboken, New Jersey
Brooklyn, New York
March 1980

Contents

Factor Analysis in Chemistry

1 Introduction

Chemists increasingly are using computers to solve chemical problems. Indeed, a new subdiscipline of chemistry called *chemometrics* is flourishing as a result of this revolutionary marriage of chemistry and computer science. Chemometrics is the utilization of mathematical and statistical methods for handling, interpreting, and predicting chemical data. Powerful methodologies have opened new vistas for chemists and have provided useful solutions for many complex chemical problems.[1,2] Factor analysis has proven to be one of the most potent techniques in the chemometric arsenal.

In Chapters 1 and 2 we provide a simplified, qualitative introduction to factor analysis. These two chapters furnish an overview of factor analysis, written especially for readers who do not wish to wrestle with mathematical details. In Chapter 1 we explore the purposes and advantages of factor analysis, while avoiding detailed methodology and mathematical derivations. For the mathematically inclined reader and for all who wish to acquire an in-depth understanding of factor analysis, a rigorous development is presented in Chapters 3 and 4.

1.1   PHILOSOPHICAL BASIS

The nature and objectives of factor analysis can be illustrated with an example from academic life. As we know, the same laboratory report will receive different grades from different professors because of variations in marking criteria. The grade assigned represents a composite of a variety of factors (i.e., subjects), such

1

as chemistry, physics, mathematics, grammar, and organization. Each factor is weighted in importance according to the personal judgment of the professor. The various grades a given report receives are due to the differences in importance conferred upon each factor by each professor.

In the technique of factor analysis, the grade is viewed as a linear sum of factors, each factor being weighted differently. Grade d_{ik} received by student i from professor k is assumed to have the form

$$d_{ik} = s_{i1}l_{1k} + s_{i2}l_{2k} + \cdots + s_{in}l_{nk} = \sum_{j=1}^{n} s_{ij}l_{jk} \tag{1.1}$$

where s_{ij} is the true *score* of student i in factor j, l_{jk} is the relative *loading* (importance) given by professor k to factor j, and the sum is taken over the n *factors* or subjects.

Factor analysis deals with a battery of grades involving a number of students and professors. The grades are arranged in a *matrix* such that each row designates a particular student and each column designates a particular professor. Such a data matrix, involving four students and three professors, is shown in (1.2). If only two factors, such as chemistry and English, were considered important in the grading, each data point could be broken down into a sum of two factors, as shown in the first equality:

$$
\begin{array}{c}
\text{professors} \\
\begin{array}{ccc} 1 & 2 & 3 \end{array}
\end{array}
$$

$$
\text{students}\begin{array}{c} 1 \\ 2 \\ 3 \\ 4 \end{array}
\begin{bmatrix}
d_{11} & d_{12} & d_{13} \\
d_{21} & d_{22} & d_{23} \\
d_{31} & d_{32} & d_{33} \\
d_{41} & d_{42} & d_{43}
\end{bmatrix}
=
\begin{bmatrix}
s_{11}l_{11} + s_{12}l_{21} & s_{11}l_{12} + s_{12}l_{22} & s_{11}l_{13} + s_{12}l_{23} \\
s_{21}l_{11} + s_{22}l_{21} & s_{21}l_{12} + s_{22}l_{22} & s_{21}l_{13} + s_{22}l_{23} \\
s_{31}l_{11} + s_{32}l_{21} & s_{31}l_{12} + s_{32}l_{22} & s_{31}l_{13} + s_{32}l_{23} \\
s_{41}l_{11} + s_{42}l_{21} & s_{41}l_{12} + s_{42}l_{22} & s_{41}l_{13} + s_{42}l_{23}
\end{bmatrix}
$$

$$
=
\text{students}\begin{array}{c} 1 \\ 2 \\ 3 \\ 4 \end{array}
\begin{bmatrix}
s_{11} & s_{12} \\
s_{21} & s_{22} \\
s_{31} & s_{32} \\
s_{41} & s_{42}
\end{bmatrix}
\begin{bmatrix}
l_{11} & l_{12} & l_{13} \\
l_{21} & l_{22} & l_{23}
\end{bmatrix}
\begin{array}{c} 1 \\ 2 \end{array}\text{factors}
\tag{1.2}
$$

The second equality in (1.2) is the result of applying standard rules of matrix multiplication. In matrix notation (1.2) becomes

$$
\underset{\text{data}}{[D]} = \underset{\text{scores}}{[S]}\;\underset{\text{loadings}}{[L]}
\tag{1.3}
$$

Here $[D]$ is the *data matrix* that consists of the grades; $[S]$ is the matrix of the students' true scores in each subject, called the *score* matrix; and $[L]$ is the matrix of importance conferred upon each subject by each professor, called the *loading matrix*.

The purpose of factor analysis, as visualized by psychologists, is to extract the student score matrix from the data matrix in order to determine the students' true abilities in each subject, in effect removing the professors' prejudices (the loadings) from the grades. Such reasoning forms the philosophical basis of factor analysis.

The form of the data matrix discussed above is analogous to many types of data matrices we encounter in chemistry, where, for example, molecules are used in place of students and chemical measurements emulate the professors. Physical observations, such as boiling points, melting points, spectral intensities, and retention values, are data analogous to grades. In a chemical problem, a row of data may concern a particular molecule and a column may concern a particular measurement. By factor analysis, we obtain a molecule score matrix, which depends solely upon the characteristics of the molecules, and a measurement loading matrix, which depends solely upon the nature of the measurements. Such a separation of the features of the molecules from the features of the measurements provides the chemist with a better insight into the true nature of the chemical phenomenon involved.

1.2   GENERALIZATIONS

We need to develop the concepts described in the previous section in a broader sense, devising a notation and terminology which can be used throughout this book in a general manner. We start with a data matrix, $[D]$, consisting of r rows and c columns:

$$[D] = \text{row designee} \begin{bmatrix} d_{11} & d_{12} \cdots d_{1c} \\ d_{21} & d_{22} \cdots d_{2c} \\ \cdot & \cdot & \cdot \\ \cdot & \cdot & \cdot \\ \cdot & \cdot & \cdot \\ d_{r1} & d_{r2} \cdots d_{rc} \end{bmatrix} \qquad (1.4)$$

column designee

The row and column headings of the matrix will be called *designees*. Each measured data point in $[D]$ is specified by a subscript denoting its row and column position in the matrix. The symbol d_{ik} represents the data point associated with the ith row designee and the kth column designee of the matrix.

Abstract Model.   The first objective of factor analysis is to obtain a mathematical, "abstract" solution wherein each point in the data matrix is expressed as a linear sum of product terms. The number of terms in the sum, n, is called the *number of factors*. Specifically, we seek solutions of the form

$$d_{ik} = \sum_{j=1}^{n} r_{ij}c_{jk} \tag{1.5}$$

In this equation the r_{ij}'s and the c_{jk}'s are called *cofactors*. For the jth factor in the sum, *row cofactor r_{ij}* is associated with the ith row designee of the data matrix and c_{jk} is the corresponding *column cofactor* associated with the kth column designee of the matrix. In classical abstract factor analysis, the row cofactors are called scores and the column cofactors are called loadings.

For data modeled by (1.5), the data matrix can be decomposed into two matrices:

$$[D] = [R]_{\text{abstract}} [C]_{\text{abstract}} \tag{1.6}$$

data   row   column

matrix matrix   matrix

where

$$[R]_{\text{abstract}} = \begin{array}{c} \\ \text{row designee} \end{array} \overset{\text{factor}}{\begin{bmatrix} r_{11} & r_{12} \cdots r_{1n} \\ r_{21} & r_{22} \cdots r_{2n} \\ \cdot & \cdot & \cdot \\ \cdot & \cdot & \cdot \\ \cdot & \cdot & \cdot \\ r_{r1} & r_{r2} \cdots r_{rm} \end{bmatrix}} \tag{1.7}$$

$$[C]_{\text{abstract}} = \begin{array}{c} \\ \text{factor} \end{array} \overset{\text{column designee}}{\begin{bmatrix} c_{11} & c_{12} \cdots c_{1c} \\ c_{21} & c_{22} \cdots c_{2c} \\ \cdot & \cdot & \cdot \\ \cdot & \cdot & \cdot \\ \cdot & \cdot & \cdot \\ c_{n1} & c_{n2} \cdots c_{nc} \end{bmatrix}} \tag{1.8}$$

Since this factor analytical solution is purely mathematical and is devoid of physical meaning, these matrices are called *abstract matrices*. The columns of $[R]_{\text{abstract}}$ are called *abstract factors*. *Row matrix $[R]_{\text{abstract}}$* contains a row for each of the r row designees and a column for each of the n factors, while *column matrix $[C]_{\text{abstract}}$* has a column for each of the c column designees and a row for each factor. The factor analytical solution isolates the row-designee factors from the column-designee factors.

Methodologies for determining the number of factors and for calculating the abstract row and column matrices are discussed in Chapters 2 through 6. Since the abstract solution should involve a physically meaningful number of factors, determination of n, the correct factor "size," is a particularly important step.

As a result of this step we obtain an estimate of the complexity of the data space, information normally lacking even for the simplest chemical problems.

Interpreting Factors. The ultimate objective of factor analysis is to develop a complete, physically meaningful model for the data. Hence the second objective of factor analysis is to convert the abstract solution into a real solution. To do this, we mathematically "transform" the abstract factors into physically significant, "real" factors. Transforming the abstract solution into a real solution is a difficult, but realizable, goal of factor analysis.

To carry out transformations, we search for an appropriate transformation matrix, $[T]$. Postmultiplying $[R]_{abstract}$ by $[T]$ and premultiplying $[C]_{abstract}$ by the inverse of the transformation matrix, $[T]^{-1}$, the data matrix in (1.6) can be expressed as

$$[D] = \{[R]_{abstract}[T]\}\{[T]^{-1}[C]_{abstract}\}$$

$$= [R]_{transformed}[C]_{transformed} \tag{1.9}$$

If the transformed solution can be shown to have physical significance, a real solution to the problem will have been found so that

$$[D] = [R]_{real}[C]_{real} \tag{1.10}$$

This equation summarizes the ultimate objective of factor analysis.

How such magical transformations can be carried out is one of the main topics of Chapters 2 through 6. Using factor analysis, we may be able to ascribe meanings to chemical data which initially appear to be exceedingly complicated because of the myriad of uncontrollable factors at play. The potential for modeling data with real factors is the most exciting feature of factor analysis.

A technique called *target testing* is especially valuable for achieving meaningful transformations. Suspected parameters (such as physical properties or structural features of molecules) can be tested individually as possible factors, and complete models of real factors can be systematically pieced together. This individual testing ability is one of the most valuable features of the target factor analysis method.

1.3 CHEMICAL EXAMPLE

To illustrate how factor analysis can be applied to chemical problems, let us consider a hypothetical data matrix, $[A]$, involving the absorbances of four different mixtures of the same absorbing components measured at five wavenumbers:

$$[A] = \begin{array}{c} \\ \text{wavenumber} \\ \begin{array}{c} 1 \\ 2 \\ 3 \\ 4 \\ 5 \end{array} \end{array} \begin{array}{cccc} & \text{mixture} & & \\ 1 & 2 & 3 & 4 \\ \begin{bmatrix} 0.371 & 0.713 & 0.219 & 0.186 \\ 0.271 & 0.515 & 0.202 & 0.174 \\ 0.229 & 0.424 & 0.241 & 0.271 \\ 0.349 & 0.641 & 0.409 & 0.428 \\ 0.182 & 0.226 & 0.229 & 0.265 \end{bmatrix} \end{array} \quad (1.11)$$

Such spectral information can be obtained from a chemical kinetics study if samples of the reaction mixture are collected at four different times during the experiment. The problem here is to determine the number of components present and to ascertain their concentrations.

According to (1.5), factor analysis will automatically furnish an abstract solution for each absorbance datum, A_{ik}, in the form

$$A_{ik} = \sum_{j=1}^{n} w_{ij}m_{jk} \quad (1.12)$$

Here w_{ij} and m_{jk} are the jth abstract row and column cofactors associated with the ith wavenumber and the kth mixture, respectively. To account for the absorbances within experimental error, n factors are included in the sum. According to (1.12), the absorbance data matrix has an abstract factor analytical solution expressed by

$$[A] = [W]_{\text{abstract}}[M]_{\text{abstract}} \quad (1.13)$$

where $[W]_{\text{abstract}}$ and $[M]_{\text{abstract}}$ are wavenumber-factor and mixture-factor matrices, respectively.

The most important feature of the abstract solution is that it reveals the number of factors responsible for the absorbance data. Ultimately, we search for an appropriate transformation matrix that will convert the abstract solution into a physically significant real solution:

$$[A] = [W]_{\text{real}}[M]_{\text{real}} \quad (1.14).$$

Going from (1.13) to (1.14) is not automatic. On the contrary, this step presents the most difficult challenge to the chemist, requiring a great deal of effort, knowledge, and intuition. If theoretical speculations can be invoked, the transformation has a better chance of being successful.

Recognizing that absorbance data for multicomponent mixtures can be approximated by Beer's law, we can interpret the factors chemically. For a mixture containing n absorbing components, Beer's law models each absorbance datum by the equation

$$A_{ik} = \sum_{j=1}^{n} \epsilon_{ij} c_{jk} \qquad (1.15)$$

Here ϵ_{ij} is the molar absorptivity per unit path length of component j at wavelength i, and c_{jk} is the molar concentration of component j in the kth mixture. Equation (1.15) involves a linear sum of products analogous to (1.12); therefore, data that obey Beer's law should have meaningful factor analytical solutions. To solve the problem completely, we must find a transformation matrix that will convert the abstract factor analytical solution into the real solution. When this is done correctly, (1.14) will take the form

$$[A] = [E]_{real}[C]_{real} \qquad (1.16)$$

Each column of the molar absorptivity matrix, $[E]_{real}$, corresponds to the absorbance of one of the pure components at the five wavenumbers, essentially tracing out the spectrum of the pure component. Each row of the molar concentration matrix, $[C]_{real}$, corresponds to the concentrations of one of the n components in each of the four mixtures.

Let us now summarize the kinds of information that might be furnished by factor analyses of absorbance data. First and quite important, determining the number of factors is tantamount to finding the total number of absorbing components in the mixtures. Second, astute transformation of the abstract factor analytical solution furnishes a good factor analytical representation of the real situation. The molar absorptivity matrix and the molar concentration matrix are the desired, physically significant transformations of $[W]_{abstract}$ and $[M]_{abstract}$. Successful transformation to $[E]_{real}$ identifies each component chemically via its spectrum. The concomitant transformation to $[C]_{real}$ furnishes the concentrations of the components in each sample mixture.

In summary, the ultimate payoff from factor analysis in this type of problem might be to determine:

1. The number of absorbing components.
2. The concentration of each component in each mixture.
3. The spectrum of each component.

The factor analytical approach is far more useful than the popular determinant method for finding the concentrations of components in multicomponent mixtures, since the spectra of all the components must be specified initially in the latter approach. By contrast, factor analysis can furnish the number of components, the concentrations, and spectral information via a purely mathematical route. Describing exactly how factor analysis can accomplish these and even other, more difficult tasks is the primary objective of this book.

1.4 ATTRIBUTES

Factor analysis often allows us to answer the most fundamental questions in a chemical problem: How many factors influence the observable? What are the natures of these factors in terms of physically significant parameters? Factor analysis enables chemists to tackle problems that in the past have had to be avoided because too many uncontrollable variables influenced the data. Factor analysis not only enables us to correlate and explain data, but also to fill in gaps in our data store.

In this section we list some of the virtues of the factor analytical approach. In particular, the following five general attributes illustrate why a chemist might want to use factor analysis:

1. *Data of great complexity can be investigated.* Factor analysis, being a method of "multivariate" analysis, can deal with many factors simultaneously. This feature is of special importance in chemistry, since interpretations of most chemical data require multivariate approaches.

2. *Large quantities of data can be analyzed.* Factor analyses can be carried out efficiently using standard factor analytical computer programs. Methods such as factor analysis are needed to properly utilize the voluminous data sets of chemistry.

3. *Many types of problems can be studied.* Factor analysis can be applied regardless of our initial lack of insight into the data. Although, ideally, factor analysis is used in conjunction with theoretical constructs, the approach can yield valuable predictions based upon empirical applications.

4. *Data can be simplified.* Matrices can be modeled concisely with a minimum of factors, and generalizations that bring out the underlying order in the data can be obtained.

5. *Factors can be interpreted in useful ways.* The nature of the factors can be clarified and deciphered, and data can be classified into specific categories. Complete physically significant models can be developed systematically, and these models can be employed to predict new data.

In general, factor analysis provides a means to attack those problems that appear to be too difficult to solve. Such problems are bountiful in chemistry, making factor analysis an ideal probe for exploration. Finding the controlling factors is akin to an engineer drilling for oil. To increase the chances for success, the engineer must use every scientific resource available; blind drilling can be extremely expensive, time-consuming, and fruitless. A great deal of scientific input and intuition is required in the factor analytical approach.

Nearly one hundred publications bear witness to the power and utility of factor analysis in chemistry. Howery[3,4] and Weiner[5] have reviewed the role of factor analysis in chemistry. Various chemical applications will be discussed in Chapters 7 through 10.

REFERENCES

1. B. R. Kowalski (Ed.), *Chemometrics: Theory and Applications,* ACS Symp. Ser. 52, American Chemical Society, Washington, D.C., 1977.
2. R. F. Hirsch (Ed.), *Statistics,* Franklin Institute Press, Philadelphia, 1978.
3. D. G. Howery, *Am. Lab.,* **8** (2), 14 (1976).
4. D. G. Howery, in R. F. Hirsch (Ed.), *ibid.,* p. 185.
5. P. H. Weiner, *Chem. Tech.,* **1977,** 321.

2 Main Steps

In this chapter we summarize the methodology of factor analysis in a qualitative manner. The chapter serves as a prelude to the rigorous development of factor analysis presented in Chapters 3 and 4.

2.1 SUMMARY OF STEPS

Factor analysis involves the following main steps: preparation, reproduction, target testing, abstract rotation, combination, and prediction. Figure 2.1 shows the sequencing of the steps and the most important information resulting from each step.

The essence of each step is as follows. In the *preparation* step the data to be factor-analyzed are selected and pretreated mathematically. The *reproduction* step, the foundation of the analysis, furnishes an abstract solution using the correct number of factors. These two steps are common to all factor analyses. Two kinds of mathematical transformations: target testing and abstract rotation, are employed to obtain more useful solutions. *Target testing* is a special technique for identifying individual, real factors. *Rotation* converts the abstract factors into easier-to-interpret abstract factors. The *combination* step furnishes

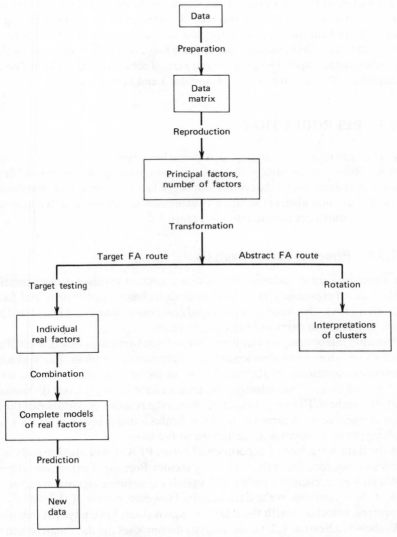

Fig. 2.1. Block diagram of the main steps in factor analysis.

complete models of real factors. Finally, new data can be calculated in the *prediction* step.

2.2   PREPARATION

The objective of the data preparation step is to obtain a data matrix best suited for factor analysis. This step involves formulating the problem, selecting the data, and mathematically pretreating the data to conform to appropriate theoretical and statistical criteria. The ultimate success or failure of factor analysis depends strongly upon the preparation step. Technical and practical details pertaining to the step are given in Chapters 3 and 6, respectively.

2.3   REPRODUCTION

The abstract reproduction step is the mathematical underpining of factor analysis. Reproduction involves two procedures: obtaining the "principal" factor solution, and determining the correct number of factors using a data reproduction method. The final abstract solution is expressed as principal factor matrices. These procedures are summarized in Figure 2.2.

2.3.1   Principal Factor Analysis

The procedure for calculating the abstract solution involves a mathematical method called *eigenanalysis*. A least-squares technique called *principal factor analysis* (PFA), also known as *principal component analysis* (PCA), is commonly employed to carry out the eigenanalysis.

Principal factor analysis yields an abstract solution consisting of a set of *abstract eigenvectors* and an associated set of *abstract eigenvalues*. Each principal eigenvector represents an abstract factor. In factor analysis we use the terms "factor" and "vector" interchangeably, since a *vector* is simply a one-dimensional array of numbers. Each eigenvalue measures the relative importance of the associated eigenvector. A large eigenvalue indicates a major factor, whereas a very small eigenvalue indicates an unimportant factor.

If the data were free of experimental error, PFA would yield exactly n eigenvectors, one for each of the controlling factors. Because of experimental error, PFA solutions in chemical problems invariably generate c eigenvectors, one for each of the c columns in the data matrix. However, only n of this set of c eigenvectors, associated with the n largest eigenvalues, have physical meaning.

As shown in Section 1.2, factor analysis decomposes the data matrix into the product of an abstract row matrix and an abstract column matrix. Standard

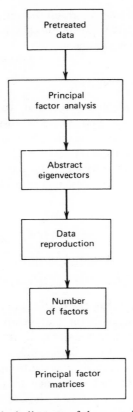

Fig. 2.2. Block diagram of the reproduction step.

factor analytical computer programs are used to calculate the complete principal factor solution: ·

$$[D] = [R]_{c,\text{PFA}}[C]_{c,\text{PFA}} \qquad (2.1)$$

Because there are c factors, the row matrix contains c columns and the column matrix contains c rows [see (1.7) and (1.8)]. The complete solution overspans the true factor space, involving more eigenvectors than are necessary.

Each row in $[C]_{c,\text{PFA}}$ of (2.1) turns out to be an abstract eigenvector. Because of the least-squares nature of PFA, the column matrix is formed using eigenvectors in decreasing order of importance. Factors are ranked according to their ability to account for variation in the data. The first row in the abstract column matrix, representing the first factor, is associated with the largest, most important eigenvalue. The cth row is least important, being associated with the smallest eigenvalue. The first factor accounts for the greatest percentage of variation in the data, the second for the next greatest percentage, and so forth, so that the complete set of c abstract factors accounts exactly for the data, including the experimental error in the measurements.

2.3.2   Number of Factors

Having calculated the complete PFA solution, we seek to discover how many of the c factors are physically important. The abstract factors can be divided into two sets: a primary set of n factors, which account for the real, measurable features of the data, and a secondary set of $c - n$ factors, which is associated entirely with experimental error. By eliminating the secondary factors from the initial solution, we "compress" the factor model to incorporate only the physically significant factors.

After compression, (2.1) becomes

$$[D] = [R]_{n,\text{PFA}}[C]_{n,\text{PFA}} \tag{2.2}$$

which for simplicity we write as

$$[D] = [R]_{\text{PFA}}[C]_{\text{PFA}} \tag{2.3}$$

Equation (2.3) concisely expresses the properly dimensioned abstract solution. All subsequent computations in factor analysis are based upon this result.

Determination of the correct factor size is the first practical dividend that we receive from the use of factor analysis. The number of real factors serves as a measure of the true complexity of the data, information seldom obtainable by other methods. If the error in the data is known, a number of methods, such as the data reproduction procedure described below, can be employed to find the number of factors. If the error in the data is not known, special mathematical techniques can be used to estimate the factor size. Empirical and theoretical methods for determining the number of real factors are explained in Chapter 4.

A stepwise, abstract reproduction procedure is often utilized to deduce the correct number of factors. Each stage of reproduction involves the following computation and comparison:

$$[R]_{j,\text{PFA}}[C]_{j,\text{PFA}} = [D]_j \overset{?}{=} [D] \tag{2.4}$$

Here $[R]_{j,\text{PFA}}$ and $[C]_{j,\text{PFA}}$ are abstract matrices based on the j most important eigenvectors, $[D]_j$ is the data matrix reproduced using the first j abstract factors, and $[D]$ is the original data matrix. In the first reproduction attempt, only the single, most important factor ($j = 1$) is used; in the second stage of reproduction, the first and second most important factors ($j = 2$) are employed simultaneously; and so forth, until all c factors are utilized together in the final reproduction.

As additional factors are incorporated in (2.4), data reproduction becomes more accurate, since, cumulatively, a greater fraction of the total variation in the data is accounted for. When the correct number of factors is employed ($j = n$), the reproduced data matrix, $[D]_n$, should equal the original data matrix within experimental error. A certain number, n, of the complete set of c factors are required to reproduce the data within experimental error. If too few factors are employed in the abstract factor analytical model, the data will not be reproduced with sufficient accuracy. If too many factors are used, the extra factors will reproduce experimental error and will therefore serve no useful purpose.

To illustrate the reproduction step, let us return to the absorbance problem described in Section 1.3. Carrying out a PFA on the hypothetical data matrix in (1.11), we find that three factors are required to reproduce the data within the built-in error of ± 0.001 absorbance unit. Thus the factor size in this problem is three, indicating that the number of absorbing components in the mixtures is three. This result is consistent with the fact that the hypothetical data were calculated from a three-component model.

When the only information required in a chemical problem is the number of factors, the preparation and reproduction steps constitute the complete factor analysis. However, in most chemical problems, the main objective is to gain insight into the nature of the factors. For this purpose the abstract solution must be transformed into a more meaningful solution.

2.4 TRANSFORMATION

Transformation of the principal factors into recognizable parameters is the most important dividend of factor analysis. As explained in Section 1.2, a transformation matrix, $[T]$, and its inverse, denoted by $[T]^{-1}$, are employed to carry out a transformation. The mathematical basis of transformation is summarized by the sequence of equations

$$[D] = [R]_{PFA}[C]_{PFA}$$
$$= \{[R]_{PFA}[T]\}\{[T]^{-1}[C]_{PFA}\}$$
$$= [R]_{transformed}[C]_{transformed} \qquad (2.5)$$

In a successful transformation, the factors in the transformed matrices will be more amenable to interpretation than will the principal factors.

Two distinctly different approaches, target testing and abstract rotation, are employed to transform the PFA solution. Target transformation is a unique method for testing potential factors one at a time. Rotation is the name given to a host of techniques for transforming PFA abstract matrices into other abstract matrices.

Because there are two major transformation methodologies, we subdivide factor analysis into two areas:

1. *Target factor analysis* (TFA), involving target testing.
2. *Abstract factor analysis* (AFA), involving PFA and rotation.

Target factor analysis, developed during the 1960s, offers special, mathematical procedures for testing physical models. We will emphasize TFA in this monograph because this approach appears to be particularly advantageous for chemical applications. New terminology and methodologies especially suited to the physical sciences will be stressed. Abstract factor analysis, rudimentary forms of which were developed during the first quarter of this century, has had a significant impact on the behavioral sciences. Much of the standard material relating to AFA will not be covered in this book since so many volumes devoted to it are available (see the Bibliography). Chemists interested in detailed discussions of AFA should consult the readable monograph of Rummel.[1]

Transformation involves four types of factors:[2]

1. Principal.
2. Rotated.
3. Typical.
4. Basic.

Principal and rotated factors are abstract factors generated by means of AFA, whereas typical and basic factors are real factors identified by means of TFA. *Principal* factors are those obtained directly from PFA solutions, as discussed in Section 2.2. By transforming the principal factors, we seek factors that will have physical meaning. Rotations of PFA matrices produce *rotated* factors in a new abstract form. Since chemists prefer models of real factors, typical and basic factors will be referred to often herein. *Typical* factors are simply rows or columns of the original data matrix which can be employed as factors. *Basic* factors, which describe the properties of the designees in the data matrix, are the most fundamental of the four types of factors.

Each of these four types of factors plays an important role in factor analysis. Principal factors serve as the basic set for all factor analytical calculations and, as stressed in Section 2.2, are used to determine the number of factors. Transforming the principal factors leads to the other three types of factors. Rotated solutions, which furnish classification information, are widely used in the behavioral sciences. Typical and basic factors, being real factors, are of special importance in the physical sciences. Using target testing, data can be modeled with a small set of "key" typical factors. By transforming to appropriate typical factors, the information in the original data matrix can be represented empirically and useful predictions can be made. Basic factors furnish insight into the underlying, physically significant influences on data. The ultimate objective of TFA is to model data with a small set of key basic factors.

2.5 TARGET TESTING

Target transformation, involving target testing, serves as a mathematical bridge between abstract and real solutions. Using target testing, we can evaluate ideas concerning the nature of the factors, and thereby develop physically significant models for data.

Regardless of the complexity of the problem, we can test potential factors individually. Tests can be associated with either the row designees or the column designees of the data matrix by focusing attention either on the row matrix or the column matrix. The procedure for target testing row-designee factors by target transformation is summarized by

$$[R]_{\text{PFA}} T = R_{\text{predicted}} \overset{?}{=} R_{\text{test}} \tag{2.6}$$

Here $[R]_{\text{PFA}}$ is the row matrix from PFA based on n factors [see (2.3)] and the other three quantities are vectors. The target transformation vector, T, results from a least-squares operation involving the PFA solution and the individual "target" being tested, designated by vector, R_{test}. If the test vector is a real factor, the predicted vector, $R_{\text{predicted}}$, obtained from (2.6) will be reasonably similar to the test vector, confirming the idea embodied in the test vector. If the test and predicted vectors are sufficiently dissimilar, the parameter tested is not a real factor, thus leading to a rejection of the tested idea.

Target testing is carried out using a special computer program developed for this purpose.[3] Mathematical criteria for evaluating the success of a target test are available (see Section 4.6).

To illustrate the power of target testing, consider the two tests shown in Table 2.1. Both tests relate to the absorbance problem discussed in Section 1.3. We use TFA to identify the components in the mixtures by searching for basic factors associated with the absorbance spectra of various pure components. For the two suspected components being tested here, absorptivities measured at the five wavenumbers constitute the two test vectors, as shown in the table. Each test is carried out separately.

Since the reproduction step indicated three factors in this problem, the two target tests each involved the three most important principal factors. The least-squares vectors, obtained from TFA via (2.6), are listed as predicted vectors in the table. In test 1 the point-by-point agreement between the test vector and the predicted vector is quite good, signifying that test component 1 is a true component. In test 2 the agreement between the test and predicted vectors is

Table 2.1   **Examples of target testing**

Wave number	Test 1 Spectrum of Potential Component 1		Test 2 Spectrum of Potential Component 2	
	Test Vector (Input)	Predicted Vector (Output)	Test Vector (Input)	Predicted Vector (Output)
1	499	504	507	863
2	501	495	983	613
3	1003	996	2031	1227
4	1502	1505	1472	1572
5	998	1001	492	1146

poor, indicating that the second test component is not present in the mixtures. These conclusions are in accord with the facts, since the hypothetical data in (1.11) were formed with the first test vector as a factor, whereas neither of the other two components in the model had spectra resembling the second test vector.

Target testing is a unique method for modeling chemical data.[4] Three important features of the target transformation approach are described below.

1. Each factor can be evaluated independently, even if a multitude of other factors simultaneously influences the data. To test the vector of interest, we do not have to identify or specify any of the other factors. This feature, in particular, distinguishes TFA from multiple regression analysis (see Section 6.2.1). The overall complexity of the data does not deter the mathematical isolation of single real factors.

2. Complete models, involving combinations of real factors, can be developed using a target-combination procedure, as discussed in Section 2.7. Factor analytical procedures can be employed to predict new data, a practical bonus of target testing.

3. Target factor analysis serves as both a theoretical tool and an empirical tool. Target testing can be utilized not only to confirm theory, but also to extend and modify theoretical models. When insight into the data is lacking, target testing can be used as a guide to search, term by term, for the best empirical model.

2.6   ROTATION

The second method for transforming factor analytical solutions involves abstract rotation. Using this approach, the column matrix obtained from PFA can be mathematically transformed into a new abstract matrix according to

$$[T]_{\text{rotation}}[C]_{\text{PFA}} = [C]_{\text{rotated}} \qquad (2.7)$$

Here $[T]_{\text{rotation}}$ is the transformation matrix required to carry out the desired rotation, $[C]_{\text{PFA}}$ is the principal factor matrix given in (2.3), and $[C]_{\text{rotated}}$ is the abstract column matrix resulting from the rotation.

There are many types of rotational criteria, each involving a different mathematical principle and yielding a different solution. A number of these methods are discussed in Chapters 16 and 17 of Rummel's monograph.[1] Packages of standard computer programs, such as the *Statistical Package for the Social Sciences,*[5] can be employed to carry out several types of rotations.

Rotated solutions are easier to interpret than are principal factor solutions. In particular, by bringing out "clusterings" in the data, rotated factors are valuable for classifying the designees in a data matrix. If a group of designees have relatively large cofactors on the same abstract factor, we say that these designees form a *cluster*. More so than in the principal factor model, specific features of the data are indicated in the rotated factors. Varimax rotation, the most popular rotation, tends to yield a few large cofactors, many small cofactors, and very few intermediate-valued cofactors, thereby emphasizing clusters of designees.

Because abstract factors are linear combinations of basic factors, it is difficult to relate rotated factors directly to real factors. Should one of the rotated factors be similar to a known basic factors, a chemist would be tempted to associate the abstract factor with the basic factor. For example, suppose that one of the abstract factors obtained from rotation of the PFA solution for the absorbance matrix in (1.11) resembled the molar absorptivity spectrum of a known compound. We might then speculate that the compound is a component in the mixtures. However, such interpretations are fraught with uncertainty since rotation cannot adequately isolate individual, basic factors. Fortunately, basic factors can be identified with greater assurance using target testing.

2.7 COMBINATION

Complete models of real factors are tested in the target-combination step. By comparing the results for different sets of real factors, the best TFA solution to a problem can be determined.

In combination TFA the data matrix is reproduced from real factors rather than from abstract factors. A row matrix, $[R]_{\text{real}}$, is formed from a selected set of n real factors which successfully passed the target test. For this combination of factors, a target transformation matrix, $[T]$, can be constructed using the target transformation vectors for each test factor from (2.7). The inverse of this matrix, denoted by $[T]^{-1}$, is then used in accordance with (1.9) to transform $[C]_{PFA}$ to a new column matrix, $[C]_{combination}$. The combination procedure can be summarized by the following equations and comparison:

$$[R]_{real}\{[T]^{-1}[C]_{PFA}\} = [R]_{real}[C]_{combination}$$
$$= [D]_{combination} \overset{?}{=} [D] \qquad (2.8)$$

If $[R]_{real}$ represents all the factors in the problem, the combination-reproduced matrix, $[D]_{combination}$, will be reasonably similar to the original data matrix, thus confirming the reliability of the tested combination of real factors.

For successful combination tests, the factors in $[R]_{real}$ are called *key* factors. Sometimes a best set of factors is dictated from theoretical principles. Otherwise, the best set of real factors can be found by examining all or selected combinations of real factors, n at a time. That set of factors which reproduces the data matrix with the least error is the best TFA model. Such models are useful for predicting new data, as explained in Section 2.8.

By testing combinations of basic factors, we try to model the data in a fundamental way, seeking an ultimate solution of the form

$$[R]_{basic}[C]_{basic} = [D] \qquad (2.9)$$

where each of the factors in $[R]_{basic}$ and $[C]_{basic}$ are basic factors. Equation (2.9) expresses the high point of TFA. Solutions of this type are the crown jewels of factor analysis. Although the effort required to develop basic solutions can be extensive, TFA can yield detailed models that cannot be obtained by other methods.

Although it may be too difficult to find key sets of basic factors in some problems, combination TFA of typical factors furnishes a routine procedure for compressing the data into a few key typical factors. Such factors can be found by target testing all or selected combinations of rows or of columns from the data matrix. With target combination, that submatrix of the original data matrix which best incorporates the factors can be determined.

2.8 PREDICTION

The target-prediction step furnishes the practical bonuses of TFA. Missing data on test vectors as well as new data rows and new data columns can be obtained using target prediction.

Data can be predicted routinely via target testing. Test vectors do not have to be complete; instead, missing values for test points can be left blank, a procedure called *free floating*. Since predicted vectors are always complete because of the nature of target testing, values for the free-floated points are predicted automatically in successful target tests. In this way, basic data that have not been measured can be estimated from TFA. For example, consider the first target test in Table 2.1. If the molar absorptivity of the test compound at one of the wavenumbers had not been known, that test point could have been free-floated in the target test, and its value would have been predicted accurately since the test vector is a real factor.

An extension of the combination-TFA procedure described in Section 2.7 can be used to add new rows and new columns to the original data matrix, thus expanding the matrix. To predict a datum associated with a new row designee, x, and an original column designee, k, we employ a modified form of the basic equation of factor analysis:

$$d_{xk}(\text{predicted}) = \sum_{j=1}^{n} r_{xj}(\text{known})c_{jk}(\text{combination}) \qquad (2.10)$$

Column cofactors c_{jk} are taken from $[C]_{\text{combination}}$, calculated in (2.8), but the values of the row cofactors r_{xj} must be known from sources independent of factor analysis. The key sets of real factors from which $[C]_{\text{combination}}$ is calculated can contain either basic or typical factors.

Table 2.2   Features of the main steps in factor analysis

Step	Purpose	Procedure	Result
1. Preparation	To obtain a matrix best suited for factor analysis	Data selection, data pretreatment	Complete data matrix in suitable form
2. Reproduction	To generate an abstract model	Principal factor analysis, stepwise abstract reproduction	Principal factor matrices, number of factors
3. Transformation			
Target testing	To evaluate test factors individually	Transformation into real factors	Identification of real factors
Rotation	To interpret abstract model	Transformation into new abstract matrices	Clusterings of data
4. Combination	To develop models from sets of real factors	Simultaneous transformation into a set of real factors	Key sets of real factors
5. Prediction	To calculate new data	Free-float missing points, employ key combination set	New target data, new data rows and columns

2.9 SYNOPSIS

The main features of each step in factor analysis are summarized in Table 2.2. Factor analysis can contribute unique answers to several of the fundamental questions which are asked in most chemical problems. Specifically, factor analysis can:

1. Furnish the number of factors that influence data.
2. Identify the real factors.
3. Detect clusterings of data.
4. Identify key sets of real factors.
5. Predict new data.

The specific dividends discussed in this chapter coupled with the general attributes discussed in Section 1.4 testify to the power of factor analysis.

REFERENCES

1. R. J. Rummel, *Applied Factor Analysis,* Northwestern University Press, Evanston, Ill., 1970.
2. R. W. Rozett and E. M. Petersen, *Anal. Chem.,* **47,** 2377 (1975).
3. E. R. Malinowski, D. G. Howery, P. H. Weiner, J. R. Soroka, P. T. Funke, R. S. Selzer, and A. Levinstone, "FACTANAL," Program 320, Quantum Chemistry Program Exchange, Indiana University, Bloomington, Ind., 1976.
4. D. G. Howery, in B. R. Kowalski (Ed.), *Chemometrics: Theory and Applications,* ACS Symp. Ser. 52, American Chemical Society, Washington, D.C., 1977, p. 73.
5. N. H. Nie, C. H. Hull, J. G. Jenkins, K. Steinbrenner, and D. H. Bent, *Statistical Package for the Social Sciences,* 2nd ed., McGraw-Hill, New York, 1975.

3 Mathematical Formulation

3.1 INTRODUCTION

In this chapter we restrict our discussion to mathematical derivations and techniques pertinent to solutions of chemical problems.[1-3] Our treatment is different from the more widely known forms of abstract factor analysis developed for psychological and sociological studies.[4-9] However, the mathematics developed is applicable in a quite general sense. Whenever possible, we correlate the present terminology with the classical work.

3.1.1   Criteria for Factor Analysis

We begin our discussion by visualizing a matrix of experimental data. We label each data point in the matrix by the symbol d_{ik}, where subscript i refers to a particular row designee and subscript k refers to a particular column designee. Factor analysis is applicable whenever a measurement can be expressed as a linear sum of product terms. More specifically, we search for solutions of the form

$$d_{ik} = \sum_{j=1}^{n} r_{ij}c_{jk} \tag{3.1}$$

where r_{ij} is the jth cofactor associated with row designee i, and c_{jk} is the jth cofactor associated with column designee k. The number of terms in the summation is n, the number of factors which adequately account for the measurement in question. This equation has the same mathematical form as that used in a multiple linear regression analysis if d_{ik} is the dependent variable. In multiple regression analysis, the c_{jk}'s are the regression coefficients and the r_{ij}'s are the independent variables. In comparison with regression analysis, the factor analytical method has many inherent advantages, which will become apparent during our discussion.

Often, we focus attention on either the row designee or the column designee. Whichever we focus attention on we call the score and its counterpart we call the loading. For example, if we call r_{ij} the score, c_{jk} is the coefficient or loading. Alternatively, if we call c_{jk} the score, r_{ij} is the coefficient or loading.

We will write $[D]$, $[R]$, and $[C]$ as matrices whose elements are d_{ik}, r_{ij}, and c_{jk}, respectively. The pair of subscripts designates the exact location of the element in the matrix: specifically, the respective row and column. From (3.1) and the definition of matrix multiplication, we may write

$$[D] = [R][C] \tag{3.2}$$

where $[D]$ is the experimental data matrix. In general, we will call $[R]$ the row matrix, since this matrix is associated with the row designees of the data matrix. Similarly, we will call $[C]$ the column matrix. In general, $[R]$ and $[C]$ are not square matrices.

Using the language developed by the pioneers of factor analysis, if we focus attention on $[R]$, we would call $[R]$ the score matrix or factor matrix and we would call $[C]$ the eigenvector matrix or loading matrix. If we focus attention on $[C]$, we would call $[C]$ the score matrix or factor matrix, and we would call $[R]$ the eigenvector matrix or loading matrix. Because this terminology can be quite confusing, we tend to favor the simplest and most specific designations: row matrix and column matrix.

3.1.2   Notation

Factor analysis involves the use of matrices, vectors, and scalars. Throughout this chapter we employ the following mathematical notation. Scalar quantities (i.e., numbers) will be represented by lowercase letters. Vectors (i.e., one-dimensional arrays of numbers) will be symbolized by capital letters. In particular, all vectors will be considered to be *column* vectors unless otherwise indicated. Row vectors will be denoted by a prime superscript. Lowercase letters will be used to designate the components of a vector. Subscripts will be used to characterize both vectors and scalars. Numbers and lowercase letters will be used as subscripts. Matrices will be designated by square brackets.

Based on this notation:

b_{ik} is a scalar

$$B_k = \begin{bmatrix} b_{1k} \\ b_{2k} \\ \vdots \\ b_{rk} \end{bmatrix} \text{ is the } k\text{th column vector}$$

$B_i' = [b_{i1} \quad b_{i2} \quad \cdots \quad b_{ic}]$ is the ith row vector

and

$$[B] = \begin{bmatrix} b_{11} & & b_{1c} \\ b_{21} & \cdots & b_{2c} \\ \vdots & & \vdots \\ b_{r1} & & b_{rc} \end{bmatrix} \text{ is a matrix}$$

3.1.3   Mathematical Synopsis

In order for us to develop a proper perspective, a brief overview of the mathematical steps of TFA will be presented in this section. The key steps are presented in Figure 3.1. The problem to be solved by factor analysis is simply this: From

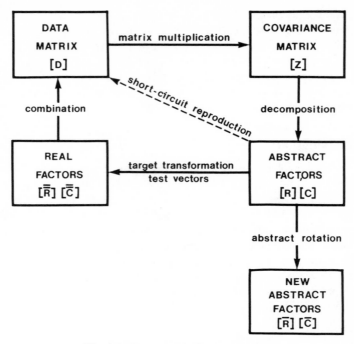

Fig. 3.1. Key steps in factor analysis.

a knowledge of $[D]$, find various sets of $[R]$ and $[C]$'s which reproduce the data in accord with (3.2). To do this we carry out the procedure described below.

The raw data are not factor-analyzed directly. Instead, they are first converted into a covariance or correlation matrix. (The difference between covariance and correlation is discussed in Section 3.2.1.) By standard mathematical techniques the covariance or correlation matrix is decomposed, via a short-circuit route, into a set of "abstract" factors which, when multiplied, reproduce the original data. These factors are called abstract because, although they have mathematical meaning, they have no real physical or chemical meaning in their present forms. Target transformation enables us to convert these factors into physically significant parameters which reproduce the experimental data.

The covariance matrix $[Z]$ is constructed by premultiplying the data matrix by its transpose:

$$[Z] = [D]^T[D] \tag{3.3}$$

This matrix is then diagonalized by finding a matrix $[Q]$ such that

$$[Q]^{-1}[Z][Q] = [\lambda_j \quad \delta_{jk}] = [\lambda] \tag{3.4}$$

Here δ_{jk} is the well-known Kronecker delta,

$$\delta_{jk} = \begin{cases} 0 & \text{if} \quad j \neq k \\ 1 & \text{if} \quad j = k \end{cases} \tag{3.5}$$

and λ_j is an eigenvalue of the set of equations

$$[Z]Q_j = \lambda_j Q_j \tag{3.6}$$

where Q_j is the jth column of $[Q]$. These columns, called *eigenvectors*, constitute a mutually orthogonal set which is usually normalized to form an orthonormal set. Hence

$$[Q]^{-1} = [Q]^T \tag{3.7}$$

We can identify $[Q]^T$ with $[C]$ by reasoning as follows:

$$\begin{aligned} [Q]^{-1}[Z][Q] &= [Q]^{-1}[D]^T[D][Q] \\ &= [Q]^T[D]^T[D][Q] \\ &= [U]^T[U] \end{aligned}$$

where

$$[U] = [D][Q] \tag{3.8}$$

Upon rearranging (3.8), we see that the data matrix can be expressed as a product of two matrices:

$$[D] = [U][Q]^T \tag{3.9}$$

By comparing this with (3.2), we see that

$$[Q]^T = [C] \tag{3.10}$$

and

$$[U] = [R] \tag{3.11}$$

We conclude that the transpose of the matrix which diagonalizes the covariance matrix represents the column matrix. Since each row of this matrix is an eigenvector, this matrix is generally called the *eigenvector matrix*. Furthermore, because the eigenvectors are orthonormal, we see that

$$[C]^{-1} = [C]^T \tag{3.12}$$

From (3.8) and (3.11) we calculate the row matrix. Having found $[R]$ and $[C]$, we can then reproduce the data matrix. Thus the short-circuit reproduction scheme (see Figure 3.1) is readily accomplished in an abstract manner.

The problem, however, is to reproduce the data within experimental error with a minimum of eigenvectors. In general, not all of the eigenvector columns of $[Q]$ are required. In Section 3.3 we show that the magnitude of the eigenvalue is a gauge of the importance of the corresponding eigenvector. Eigenvectors associated with the largest eigenvalues are most important and eigenvectors associated with the smallest eigenvalues are least important. In practice, the least important eigenvectors are negligible and are dropped from the analysis. In fact, their retention would simply regenerate experimental error. Unnecessary reproduction of error by the inclusion of too many eigenvectors is contrary to the original intent of factor analysis, which is to reproduce the data within experimental error using the minimum number of eigenvectors. Nothing is gained by retaining an excessive number of eigenvectors. The dropping of unwanted eigenvectors is called *factor compression*.

As a first trial we recalculate the data matrix using only the most important eigenvector, Q_1, which is associated with the largest eigenvalue, λ_1. In accord with (3.10), $C_1^T = Q_1$. Premultipling C_1^T by R_1, outer-product-wise, we obtain the recalculated data matrix. We then examine the recalculated data matrix to see whether or not it agrees with the original data matrix. If the agreement is within experimental error, we conclude that only one factor is important. If not, we proceed to make a second trial using the two most important eigenvectors (associated with the two largest eigenvalues, λ_1, and λ_2), letting $[R] = [R_1 \quad R_2]$ and $[C] = [C_1 \quad C_2]^T$. Again we ascertain whether or not the reproduced data adequately represent the original data. We continue adding eigenvectors associated with the largest eigenvalues, sequentially, until we satisfactorily reproduce the data:

$$[R_1 \quad R_2 \quad \cdots \quad R_n] \begin{bmatrix} C_1' \\ C_2' \\ \vdots \\ C_n' \end{bmatrix} = [R^\ddagger][C^\ddagger] = [D^\ddagger][\simeq [D] \qquad (3.13)$$

We use the superscript dagger to indicate that only n important (primary) eigenvectors are employed in constructing the row-factor and column-factor matrices. $[D^\ddagger]$ is the reproduced data matrix, using n eigenvectors. The minimum number of eigenvectors, n, required to reproduce the data within experimental error represents the number of factors involved. This number also represents the "dimensionality" (rank or size) of the factor space.

At this stage of the FA scheme, the short-circuit reproduction (see Figure 3.1) is properly completed. However, in their present forms the row and column matrices represent mathematical solutions to the problem and are not recognizable in terms of any known chemical or physical quantities. Nevertheless, in their abstract mathematical forms, these factors contain useful information. Examples of the utility of the short-circuit AFA reproduction scheme are presented in detail later in the book.

Most often the chemist is not interested in the abstract factors produced by AFA but is interested in finding the physically significant factors that produce the data. Factor analysis can be used in a powerful and unique fashion for such purposes. This is achieved by attempting to transform the abstract eigenvectors into vectors that have physical meaning. A method called *target transformation* allows us to test and develop theoretical and empirical hypotheses concerning the fundamental nature of the variables at play.

Mathematical transformation is accomplished by performing the following matrix multiplication:

$$[\overline{R}] = [R^{\ddagger}][T] \tag{3.14}$$

Here $[T]$ is the appropriate transformation matrix and $[\overline{R}]$ is the row matrix in the new coordinate system. The corresponding column matrix (the eigenvector matrix) is obtained as follows:

$$[\overline{C}] = [T]^{-1}[C^{\ddagger}] \tag{3.15}$$

This is true since

$$
\begin{aligned}
[D^{\ddagger}] &= [R^{\ddagger}][C^{\ddagger}] \\
&= [R^{\ddagger}][T][T]^{-1}[C^{\ddagger}] \\
&= [\overline{R}][\overline{C}]
\end{aligned}
\tag{3.16}
$$

Transformation matrices can be obtained by many different methods. The transformation of special interest to chemists is target transformation. This method consists of finding a set of transformation vectors by means of a least-squares method. Each transformation vector is a column of the transformation matrix. Target transformation refers to the procedure of testing a single physical or structural concept. By means of target transformation we can test theoretical concepts and obtain physically significant parameters which reproduce the data (see Figure 3.1). Details of this feature of factor analysis are presented in Section 3.4.

3.2 PRELIMINARY CONSIDERATIONS

3.2.1 Constructing the Covariance Matrix

The covariance matrix is obtained by premultiplying the data matrix by its transpose:

$$[Z] = [D]^{T}[D] \tag{3.17}$$

Although the data matrix is not necessarily square, the covariance matrix is square. If the data matrix has r rows and c columns, the covariance matrix will be of size $c \times c$. (More conventionally, the covariance matrix is constructed by postmultiplying the data matrix by its transpose:

$$[Z] = [D][D]^T$$

In this case the covariance matrix will be of size $r \times r$. Either definition can be used and both will produce compatible results in the final analysis.

These definitions automatically place a latent statistical bias on the analysis. Each data point is inherently weighted in proportion to its absolute value. Large data points are given more statistical importance than are smaller values. Such bias is desirable when all the measurements are made relative to an absolute standard and all data points bear the same absolute uncertainty.

In certain problems we may wish to give equal statistical weight to each column of data. This is accomplished by normalizing each column of data. Normalization is carried out by dividing every element in a given data column by the square root of the sum of the squares of all the elements in the column. (This is not the standard statistical definition of normalization, which involves normalization relative to the mean.) Normalization is warranted when the experimental error is directly proportional to the magnitude of the measurement, when each column of data involves a different property having significantly different orders of magnitude, or when the reference standard is arbitrary, thus prejudicing the magnitudes of the data points. This procedure is known as *correlation* and the resulting correlation matrix is denoted by $[Z]_N$.

A more thorough discussion of covariance and correlation, as well as other interesting aspects of the statistical nature of factor analysis, can be found in the text by Rummel.[4] In our mathematical development, it is not necessary for us to constantly make a distinction between the original data matrix and the normalized data matrix because such a distinction is not critical to the overall understanding of the mathematical steps that follow. Upon completion of the analysis we realize that, if normalized data are used in the analysis, a normalized data matrix will emerge from the computations.

3.2.2 Effect of Transposing the Data Matrix

If the size of the data matrix is $r \times c$ (r rows and c columns), the size of the covariance matrix will be $c \times c$. However, if the data matrix is transposed (i.e., rows and columns are interchanged) prior to forming the covariance matrix, the covariance matrix will be of size $r \times r$. Carrying out the factor analysis on either of these matrices will yield the same conclusion concerning the dimensionality of the factor space.

If covariance is used, then exactly the same eigenvectors and eigenvalues will emerge from the factor analysis regardless of whether the data matrix or its transpose is involved. This is not true if the correlation matrix is used instead of covariance.

3.2.3 Data Preprocessing

Four common methods of data processing have been used in factor analysis studies: (1) covariance about the origin, (2) covariance about the mean, (3) correlation about the origin, and (4) correlation about the mean. The relationship among these different methods depends upon a simple linear transformation:

$$[D]^{\#} = [D][A] + [B] \tag{3.18}$$

Here $[D]^{\#}$ rather than $[D]$ represents the processed data which are subjected to factor analysis. The four methods differ in the definitions of $[A]$ and $[B]$. $[A]$ is a diagonal matrix that adjusts the overall magnitude of each data column, consisting of diagonal elements a_{jj} only. $[B]$ is a matrix in which all the b_{ij} elements in any one column are identical. This matrix shifts the origin of the factor space.

1. For covariance about the origin, C_o:

$$a_{jj} = 1$$
$$b_{ij} = 0 \tag{3.19}$$

2. For covariance about the mean, C_m:

$$a_{jj} = 1$$
$$b_{ij} = -\hat{d}_j \tag{3.20}$$

3. For correlation about the origin, R_o:

$$a_{jj} = \left(\sum_{i=1}^{r} d_{ij}^2 \right)^{-1/2}$$
$$b_{ij} = 0 \tag{3.21}$$

4. For correlation about the mean, R_m:

$$a_{jj} = \left[\sum_{i=1}^{r} (d_{ij} - \hat{d}_j)^2 \right]^{-1/2}$$
$$b_{ij} = -\hat{d}_j a_{jj} \tag{3.22}$$

Here $\hat{d}_j$ is the average value of the experimental data points of the jth column of the raw data matrix, and r is the total number of points in a column of data.

Rozett and Petersen[10] discussed the advantages and disadvantages of these four methods. In chemistry, covariance about the origin is preferred because it preserves the origin of the factor space, the relative lengths of factor axes, and the relative error—a most desirable situation. By using covariance or correlation about the mean, we lose information concerning the zero point of the experimental scale. The addition of more data points will shift the origin of the factor space. By using correlation (about the origin or about the mean), we lose information concerning the relative size and relative error associated with the various data columns.

Other data preprocessing methods include algebraic transformations of each data point, such as logarithms, exponentials, squares, square roots, and reciprocals. Preferably, these transformations should be based upon some scientific principle. The success of factor analysis often depends upon the judicious choice of the data preprocessing scheme, an aspect that cannot be overemphasized.

3.2.4   Vector Interpretation

An insight into the overall operational details of factor analysis can be obtained from a vector viewpoint. With this perspective in mind we consider the columns of the data matrix to be vectors. The elements of the covariance matrix are generated by taking dot products (scalar products) of every pair of columns in the data matrix [see (3.17)]. On the other hand, when forming the correlation matrix, we normalize each column of data before taking the scalar products. Each element of the correlation matrix represents the cosine of the angle between the two respective data column vectors. The diagonal elements are unity since they are formed by taking dot products of the vectors onto themselves.

If n eigenvectors are needed to reproduce the data matrix, all the column vectors will lie in n space, requiring n orthogonal reference axes. This is best understood by studying a specific example. For this purpose let us consider a normalized data matrix, consisting of four data columns (D_1, D_2, D_3, and D_4), generated from two factors. From this matrix (not shown here), the following correlation matrix is formed:

$$[Z]_N = \begin{array}{c} \\ D_1 \\ D_2 \\ D_3 \\ D_4 \end{array} \begin{array}{cccc} D_1 & D_2 & D_3 & D_4 \\ \left[\begin{array}{cccc} 1.00000 & 0.06976 & -0.58779 & 0.80902 \\ 0.06976 & 1.00000 & 0.76604 & 0.64279 \\ -0.58779 & 0.76604 & 1.00000 & 0.00000 \\ 0.80902 & 0.64279 & 0.00000 & 1.00000 \end{array} \right] \end{array} \quad (3.23)$$

The elements of this matrix are the cosines of the angles between the data column vectors. From the numerical values given in (3.23), we can construct the graphical representation illustrated in Figure 3.2. We find that all four data

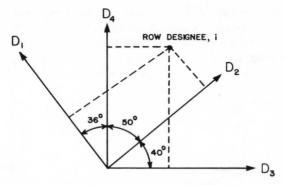

Fig. 3.2. Vector relationships of data column vectors used in forming the correlation matrix of (3.23). Data points are obtained by the perpendicular projections of a row-designee point onto the respective column-designee vectors, D_1, D_2, D_3, and D_4.

vectors lie in a common plane. The problem is two-dimensional (i.e., only two factors are involved).

Each vector axis in Figure 3.2 corresponds to a column designee of the normalized data matrix. Each row designee of the normalized data matrix is represented by a point in the two-dimensional plane. A typical row-designee point is portrayed in the figure. The value of a data point associated with a given row and column is obtained by first drawing a line through the row-designee point, perpendicular to the appropriate column vector, and then reading the distance along the vector from the origin to the intersection. These projections are the normalized data values, since the vector axes represent normalized data columns. In a more general sense, the value of a point projected onto any axis is called the *projection* or *factor component.* In classical factor analysis these projections are called "scores."

If one of the vectors in the figure did not lie in the plane, the space would be three-dimensional. Three factors would be required to account for the data. Three axes would be required to locate the data points in the factor space. The rank of the correlation matrix would be three. The choice of reference axes is not totally arbitrary. If vector 1 projected out of the plane, for example, we could use vector 1 and any two of the other three vectors as reference axes. We could not use vectors 2, 3, and 4 as reference axes, since they lie in a common plane and do not span the three-dimensional space.

In many problems the factor space has more than three dimensions. It is impossible for us to sketch such multidimensional situations on two-dimensional graph paper. However, it is possible to extract all of the necessary information with the aid of a computer. Using factor analysis we can determine the exact dimensions of the factor space. The eigenvectors that emerge from factor analysis span the factor space but do not coincide with the data vectors; they merely define the factor space in which all the experimental data points coexist.

When the correlation matrix of (3.23) is subjected to the decomposition step (see Figure 3.1), two mutually orthonormal eigenvectors, C_1 and C_2, and their associated eigenvalues, λ_1 and λ_2, emerge:

$$C_1 = \begin{bmatrix} c_{11} \\ c_{12} \\ c_{13} \\ c_{14} \end{bmatrix} = \begin{bmatrix} 0.5084 \\ 0.5084 \\ 0.0847 \\ 0.6899 \end{bmatrix} \quad \text{and} \quad C_2 = \begin{bmatrix} c_{21} \\ c_{22} \\ c_{23} \\ c_{24} \end{bmatrix} = \begin{bmatrix} 0.4909 \\ -0.4909 \\ -0.7144 \\ +0.0877 \end{bmatrix}$$

$$\lambda_1 = 2.070 \quad \text{and} \quad \lambda_2 = 1.930$$

The c_{jk} coefficients that define the eigenvectors also measure the importance of each eigenvector on each data column. In general, it can be shown that each of the four data-column vectors can be expressed in terms of the basic eigenvectors in the following way:

$$D_k = \sum_{j=1}^{n} \sqrt{\lambda_j}\, c_{jk} C_j \tag{3.24}$$

Accordingly, for our problem, the four data-column vectors may be expressed in terms of the two eigenvectors:

$$\begin{aligned} D_1 &= 0.7314C_1 + 0.6819C_2 \\ D_2 &= 0.7314C_1 - 0.6819C_2 \\ D_3 &= 0.1218C_1 - 0.9926C_2 \\ D_4 &= 0.9926C_1 + 0.1218C_2 \end{aligned} \tag{3.25}$$

The validity of these equations can be verified by taking dot products between the vectors and comparing the results with the correlation matrix given in (3.23).

In classical factor analysis the $\sqrt{\lambda_k}\, c_{jk}$ coefficients are better known as "factor loadings" or simply "loadings." The loadings are a measure of the relative importance of each eigenvector on each of the data-column vectors. They are similar to the "weightings" in regression analysis.

The geometrical relationship between the two eigenvectors and the four data column vectors can be obtained by taking dot products between the eigenvectors and the data vectors as expressed in (3.25). Recall that the dot product equals the cosine of the angle between the two vectors. The relationships so obtained are illustrated in Figure 3.3. We make the following observation. From the tips of each data vector, we draw lines parallel to each of the two eigenvectors. These lines intersect the eigenvectors. The distance measured on an eigenvector from the origin to the point of intersection represents the loading of the eigenvector

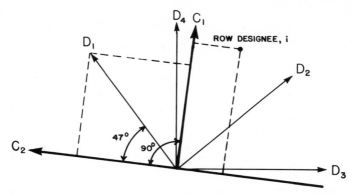

Fig. 3.3. Geometrical location of eigenvectors C_1 and C_2 resulting from factor analysis. Dashed lines show the factor loadings of C_1 and C_2 on D_1, and the scores (projections) of row-designee i on factors C_1 and C_2.

on the respective data-column vector. These loadings are the coefficients in (3.25).

Next, let us examine the nature of the points that lie in the factor space. Recall that each row designee of the data matrix corresponds to a point in the factor space. Since there are r row designees, there are r points. For our example, all of these points lie in the plane. The projection of a point onto a basis axis gives its score on the axis. Figure 3.3 shows the projection of one such point, i, projected on C_1 and C_2. These projections are the values for elements r_{i1} and r_{i2}, respectively, in the row matrix. The row matrix, $[R]$, is composed of all such projections onto the eigenvector axes.

Because there are only two basic axes, all the data information associated with this problem can be compressed and stored on two factors. For example, to locate a new row designee point in the factor plane, only two measurements involving only two column designees are required. The projections of this point onto the two remaining column designees yield predictions for the behavior of the new row designee.

Of course, the example given here is superficially small. In actual practice, the data matrix may consist of many more data columns. If the data contained 25 columns and the factor space were four-dimensional, for example, then four measurements would be required to locate a row-designee point in the factor space. The projections of this single point onto the remaining 21 column-designee axes would yield 21 predicted values. This is interesting when we realize that these predictions are made in an abstract manner requiring no information about the true controlling factors. Considering the enormous number of chemicals that could comprise the row designees, this feature of AFA offers useful possibilities to the chemist.

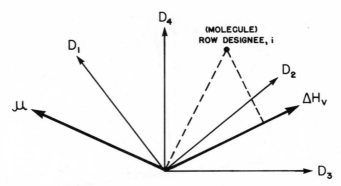

Fig. 3.4. Two real factors (dipole moment, μ, and heat of vaporization, ΔH_v) responsible for four different data columns, D_1, D_2, D_3, and D_4.

Equations (3.25) represent abstract solutions to the problem because the eigenvectors, C_1 and C_2, are devoid of any real physical or chemical meaning. For many chemical problems such solutions are sufficient and the factor analytical study is terminated at this point. However, we search for the real axes because there are many advantages to be gained if we succeed, as discussed in latter chapters. The real axes can often be found by using target transformation, described in Section 3.4. Target transformation allows us to test our ideas of what the real factors may be. Figure 3.4 illustrates the location of two hypothetical real factors, the dipole moment, μ, and the heat of vaporization, ΔH_v. The projection of the row-designee points associated with various chemicals onto these two real axes yields predictions for the dipole moments and heats of vaporization. For the molecule shown in Figure 3.4, the dipole moment is predicted to be zero and the heat of vaporization is given by the distance along the ΔH_v axis. By identifying the real axes, we automatically arrive at a new method for predicting values associated with the real axes. This is another valuable feature of TFA.

3.3   FACTOR COMPRESSION AND DATA REPRODUCTION

3.3.1   Decomposition

The rank of the covariance or correlation matrix, and hence the dimensionality of the factor space, is determined mathematically by decomposing the covariance or correlation matrix into a set of eigenvectors. If the data were pure, exactly n eigenvectors would emerge. However, because of experimental error, the number of eigenvectors will equal either r, the number of rows, or c, the number of columns in the data matrix, whichever is smaller. For convenience we will assume throughout this chapter that c is smaller than r and therefore that c eigenvectors emerge. In order to understand the theoretical principles involved in the decomposition, it is convenient to introduce a basis into the factor scheme. We will let $U_1, U_2, \ldots, U_c$ be the basis set of unit vectors that spans the factor space, [i.e., $U_1 = (1, 0, 0, \ldots, 0)$, $U_2 = (0, 1, 0, \ldots, 0)$, $U_3 = (0, 0, 1, \ldots, 0)$, etc.]. These vectors are orthonormal, so that

$$U_j U_k = \delta_{jk} \tag{3.26}$$

where δ_{jk} is the Kroenecker delta.

A data point can be looked upon as a vector in factor space. Hence instead of (3.1), a better representation is the following:

$$D_{ik} = \sum_{j=1}^{c} r_{ij} U_j c_{jk} \tag{3.27}$$

where D_{ik} is the data point.

The z_{ab} element of the covariance matrix is given by

$$z_{ab} = \sum_{i=1}^{r} D_{ia} \cdot D_{ib} = \sum_{i=1}^{r} \left(\sum_{j=1}^{c} r_{ij} U_j c_{ja} \right) \left(\sum_{k=1}^{c} r_{ik} U_k c_{kb} \right)$$

$$= \sum_{i=1}^{r} \sum_{j=1}^{c} r_{ij}^2 c_{ja} c_{jb} \tag{3.28}$$

The last step in (3.28) is true because $U_j U_k = \delta_{jk}$. From this equation we see that the entire covariance matrix can be decomposed into a sum of dyads times their corresponding eigenvalues:

$$[Z] = \sum_{i=1}^{n} r_{i1}^2 \begin{bmatrix} c_{11} \\ c_{12} \\ \vdots \\ c_{1c} \end{bmatrix} [c_{11} \quad c_{12} \quad \cdots \quad c_{1c}]$$

$$+ \sum_{i=1}^{r} r_{i2}^2 \begin{bmatrix} c_{21} \\ c_{22} \\ \vdots \\ c_{2c} \end{bmatrix} [c_{21} \quad c_{22} \quad \cdots \quad c_{2c}]$$

$$+ \cdots + \sum_{i=1}^{r} r_{ic}^2 \begin{bmatrix} c_{c1} \\ c_{c2} \\ \vdots \\ c_{cc} \end{bmatrix} [c_{c1} \quad c_{c2} \quad \cdots \quad c_{cc}] \tag{3.29}$$

In simpler notation (3.29) can be written as

$$[Z] = \lambda_1 C_1 C_1' + \lambda_2 C_2 C_2' + \cdots + \lambda_c C_c C_c' \tag{3.30}$$

in which

$$\lambda_j = \sum_{i=1}^{r} r_{ij}^2 \tag{3.31}$$

and

$$C_j' = [c_{j1} \quad c_{j2} \quad \cdots \quad c_{jc}] \tag{3.32}$$

In (3.31) λ_j is the eigenvalue associated with eigenvector C_j. We note that the eigenvalue is a sum of the squares of the projections of each row designee onto the appropriate eigenvector and hence represents the relative importance of the unit vector.

Since there are c columns of data, c eigenvectors will emerge from the decomposition process. As shown in Chapter 4, only n of these (i.e., those associated with the largest eigenvalues) are required to account for the data. The remaining $(c - n)$ eigenvectors are the result of experimental error. If there were no error in the data, exactly n eigenvectors would emerge. The number of important eigenvectors represents the dimensions of the factor space. Mathematically, this means that the true rank of the data matrix is equal to the dimensions of the factor space.

3.3.2   Principal Factor Analysis

There are a variety of mathematical ways to decompose the covariance matrix. Principal factor analysis (PFA), also known as principal component analysis (PCA), is by far the most widely used method. In PFA the eigenvectors are consecutively calculated so as to minimize the residual error in each step. Thus each successive eigenvector accounts for a maximum of the variation in the data. In this section we will trace through the derivation in order to illustrate the philosophical and mathematical principles involved.

To locate the abstract eigenvectors, the following mathematical reasoning is involved. The principal eigenvectors constitute an optimized, mutually orthogonal coordinate system. Each successive eigenvector accounts for the maximum possible variance in the data. [A formal definition of variance is given by (4.67).] The eigenvector associated with the largest, most important eigenvalue is oriented in the factor space so as to account in a least-square sense for the greatest possible variance in the data. This primary vector passes through the greatest concentration of data points. The first eigenvector defines the best one-factor model for the data. Typically, the first factor accounts for a major fraction of the variance in the data. The first abstract factor represents a sort of common factor averaged over all the designees.

This calculational procedure is continued stepwise until all c axes have been located. To find each succeeding axis, two conditions are imposed: (1) as much variance as possible is accounted for by each factor, and (2) the newest axis is mutually orthogonal to the set of axes already located. Subsequest eigenvectors account for deviations from the average behavior defined by the first eigenvector. Each successive factor is responsible for a smaller fraction of the total variance in the data. The second principal axis is orthogonal (at right angles) to the first eigenvector. This factor points in the direction that will account for as much as possible of the variance not accounted for by the first factor. The first two factors define a plane passing through the greatest concentration of data points. Points in the plane are specified by two coordinate axes in a two-factor model.

Eigenvectors associated with successively smaller eigenvalues account more and more for relatively "unique" behavior associated with only a few designees or even a single designee. In chemical problems, the smaller eigenvalues account for unique behavior and finally for experimental error only. The cth eigenvector is situated in factor space so as to account for the last bit of experimental error. The complete set of c eigenvectors accounts exactly for every bit of the data, including the experimental error.

To keep track of the factors, we will use parentheses to indicate the number of factors being considered. For example, $d_{ik}(m)$ is the reproduced data point in the ith row and kth column calculated from the first m principal factors. Hence we write

$$d_{ik}(m) = \sum_{j=1}^{m} r_{ij}c_{jk} \qquad (3.33)$$

where the sum is taken over the first m principal factors. To obtain the first principal component (factor), we proceed as follows. First, we define the residual error, $e_{ik}(1)$, to be the difference between the experimental data point, d_{ik}, and the reproduced data point, $d_{ik}(1)$, using one factor (the first principal factor):

$$e_{ik}(1) = d_{ik} - d_{ik}(1) \qquad (3.34)$$

From (3.33) we see that

$$e_{ik}(1) = d_{ik} - r_{i1}c_{1k} \qquad (3.35)$$

To minimize the residual error, we apply the method of least squares. Accordingly, we take the derivative of the square of each residual error with respect to either the row cofactors or the column cofactors, depending upon which type of designee we wish to focus attention on. If we take the derivatives with respect to the row cofactors, we sum over the c columns. On the other hand, if we take the derivatives with respect to the column cofactors, we sum over the r rows. Choosing to focus our attention on the row designees, we use the latter procedure, obtaining

$$\sum_{i=1}^{r} \frac{de_{ik}^2(1)}{dc_{1k}} = 2c_{1k} \sum_{i=1}^{r} r_{i1}^2 - 2 \sum_{i=1}^{r} r_{i1}d_{ik} \qquad (3.36)$$

Following the least-squares principle, we set this sum equal to zero, finding

$$\sum_{i=1}^{r} r_{i1}d_{ik} = c_{1k} \sum_{i=1}^{r} r_{i1}^2 \qquad (3.37)$$

Since k varies from 1 to c, there are c equations of this form, which in matrix notation can be expressed as follows:

$$R_1'[D] = C_1'R_1'R_1 \qquad (3.38)$$

We now define λ_1, consistent with (3.31), by

$$\lambda_1 = R_1'R_1 = \sum_{i=1}^{r} r_{i1}^2 \qquad (3.39)$$

Inserting this into (3.38) and taking the transpose, we find that

$$[D]^T r_1 = \lambda_1 C_1 \qquad (3.40)$$

According to (3.13), the complete data matrix can be written as

$$[D] = R_1 C_1' + R_2 C_2' + \cdots + R_c C_c' \qquad (3.41)$$

Since the sum is taken over all c eigenvectors, this equation concerns $[D]$, the complete data matrix, including experimental error, and not $[D^\ddagger]$, which involves summing over only n principal eigenvectors. Postmultiplying (3.41) by C_1 and setting $C_i'C_j = \delta_{ij}$, so that the eigenvectors are orthornormal, we obtain

$$[D]C_1 = R_1 \qquad (3.42)$$

Inserting (3.42) into (3.40), we conclude that

$$[D]^T[D]C_1 = \lambda_1 C_1 \qquad (3.43)$$

Upon recalling the definition of the covariance matrix [see (3.17)], we now write

$$[Z]C_1 = \lambda_1 C_1 \qquad (3.44)$$

As described in Section 3.3.3, this expression can be used to calculate the first principal eigenvectors, C_1, and its associated eigenvalue, λ_1.

To obtain the second principal component, we consider the second residual error,

$$e_{ik}(2) = d_{ik} - d_{ik}(2) \qquad (3.45)$$

which, from the definition given in (3.33), can be expressed as follows:

$$e_{ik}(2) = e_{ik}(1) - r_{i2}c_{2k} \tag{3.46}$$

To minimize the error in the second principal component we apply the method of least squares to $e_{ik}(2)$ while keeping $e_{ik}(1)$ constant. Thus we find an expression analogous to (3.36):

$$\sum_{i=1}^{r} \frac{de_{ik}^2(2)}{dc_{2k}} = 2c_{2k} \sum_{i=1}^{r} r_{i2}^2 - 2 \sum_{i=1}^{r} r_{i2}e_{ik}(1) \tag{3.47}$$

To minimize the error, this sum is set equal to zero, giving

$$\sum_{i=1}^{r} r_{i2}e_{ik}(1) = c_{2k} \sum_{i=1}^{r} r_{i2}^2 \tag{3.48}$$

There are c equations of this type, which in matrix notation take the following form:

$$R_2'[E]_1 = C_2'R_2'R_2 \tag{3.49}$$

where $[E]_1$ is an $r \times c$ error matrix composed of the first residual errors. Now λ_2 is defined as

$$\lambda_2 = R_2'R_2 = \sum_{i=1}^{r} r_{i2}^2 \tag{3.50}$$

From (3.49) and (3.50) we learn that

$$[E]_1R_2 = \lambda_2C_2 \tag{3.51}$$

Matrix $[E]_1$, however, can be written as

$$[E]_1 = [D] - R_1C_1' = R_2C_2' + R_3C_3' + \cdots + R_cC_c' \tag{3.52}$$

Postmultiplying (3.52) by C_2 and recalling that the eigenvectors are to be orthonormal, we obtain

$$[E]_1C_2 = R_2 \tag{3.53}$$

Inserting this equation into (3.51) gives

$$[E]_1^T[E]_1C_2 = \lambda_2C_2 \tag{3.54}$$

From (3.38), (3.39), and (3.52) we can show that

$$[E]_1^T[E]_1 = [D]^T[D] - \lambda_1C_1C_1' \tag{3.55}$$

The first residual matrix is defined as

$$[\mathcal{R}]_1 = [Z] - \lambda_1C_1C_1' \tag{3.56}$$

Hence we conclude from (3.54), (3.55), and (3.56) that

$$[\mathcal{R}]_1C_2 = \lambda_2C_2 \tag{3.57}$$

This expression, as described in the Section 3.3.3, can be used to calculate the numerical values of the second principal eigenvector, C_2, and its associated eigenvalue, λ_2.

To obtain the third principal component we apply the method of least squares to $e_{ik}(3)$. In this case we obtain

$$[\mathcal{R}]_2 C_3 = \lambda_3 C_3 \qquad (3.58)$$

where the second residual matrix is defined as

$$[\mathcal{R}]_2 = [Z] - \lambda_1 C_1 C_1' - \lambda_2 C_2 C_2' \qquad (3.59)$$

We continue in this fashion to successively extract the remaining eigenvectors. In general, we find that

$$[\mathcal{R}]_m C_{m+1} = \lambda_{m+1} C_{m+1} \qquad (3.60)$$

where

$$[\mathcal{R}]_m = [Z] - \sum_{j=1}^{m} \lambda_j C_j C_j' \qquad (3.61)$$

3.3.3   Example of Decomposition Procedure

The decomposition step of factor analysis is a time-consuming process which, from a practical point of view, is best carried out with the aid of a computer. There are many different mathematical strategies that can be used to accomplish this task. The strategies developed for computer computations differ considerably from those developed for hand computations. The iteration method described below is selected for pedagogical reasons.

Initially, numerical values for the elements of C_1 are chosen at random. It is desirable, however, to normalize the eigenvectors at each stage so that we deal with an orthonormal set. According to (3.44), multiplication of this trial vector by $[Z]$ yields $\lambda_1 C_1$, where λ_1 is the normalization constant and C_1 is the normalized eigenvector. The resulting values for C_1 represent a new and better approximation for C_1. Upon multiplication of the new approximate C_1 by $[Z]$, we again obtain $\lambda_1 C_1$; and again we obtain better values for the elements of C_1. This iteration procedure is continued until the elements of C_1 converge to constant values and (3.44) is obeyed.

To obtain C_2 we proceed as follows. First, $\lambda_1 C_1 C_1'$ is computed and subtracted from the covariance matrix yielding the first residual matrix, $[\mathcal{R}]_1$ [see (3.56)]. Examining (3.57), we see that an iteration process will yield C_2 and λ_2. To start the iteration we first arbitrarily choose numerical values for the elements of C_2. These values are multiplied by $[\mathcal{R}]_1$ as shown in (3.57), yielding a new normalized set of values for C_2. The new values are then multiplied by $[\mathcal{R}]_1$ and the iteration is carried out until the elements of C_2 converge to a constant set of values and (3.57) is satisfied.

To obtain C_3 we first calculate the second residual matrix, $[\mathcal{R}]_2$, by subtracting $\lambda_2 C_2 C_2'$ from the first residual matrix. Again we arbitrarily choose values for C_3 and apply (3.58), iterating until the equation is satisfied.

By calculating the third, fourth, and ... residual matrices, we continue to apply the iteration method to obtain C_4, C_5, ..., and C_c and their associated eigenvalues. A detailed numerical example of this procedure is presented in Chapter 5.

This method, principal factor analysis (PFA), yields a unique set of mutually orthonormal eigenvectors which represents the coordinate axes of the data space. The first eigenvector that emerges from the iteration is associated with the largest eigenvalue and accounts for most of the variance of the data. This vector is oriented in a direction that maximizes the projections of the data points onto this axis. The second eigenvector is orthogonal to the first and is also oriented so that the projections on this axis are maximized. In fact, each eigenvector that emerges from the iteration is orthogonal to all the previous eigenvectors and is oriented in the direction that maximizes the sum of squares of all projections on the axis.

If there were no error in the data, exactly n eigenvectors would emerge. However, in chemistry, perfect data are impossible to obtain. As dictated by mathematics, c eigenvectors (c being equal to the number of data columns) will always emerge from the decomposition. However, only the first n largest eigenvectors are required to account for the data within experimental error. The remaining $c - n$ eigenvectors merely account for experimental error and should be deleted from further consideration. Methods for deciding the value of n are discussed in Section 4.3.

3.3.4 Calculating the Column Matrix

The decomposition of the covariance matrix is tantamount to matrix *diagonalization,* the latter terminology being more commonly used. To show that this is true, we make the following observations. Equation (3.30) can be expressed as a product of matrices:

$$[Z] = [C]^T[\lambda][C] \qquad (3.62)$$

where

$$[C] = \begin{bmatrix} C_1' \\ C_2' \\ \vdots \\ C_c' \end{bmatrix} = [C_1 \quad C_2 \quad \cdots \quad C_c]^T \qquad (3.63)$$

Here we see that the eigenvectors, which result from the iteration process, constitute the respective rows of the column matrix, $[C]$. Notice that we consider here the complete set of c eigenvectors which account for all the data, including experimental error. Because the rows of $[C]$ are orthonormal, its transpose equals its inverse:

$$[C]^T = [C]^{-1} \tag{3.64}$$

Hence (3.62) can be rearranged to read

$$[C][Z][C]^{-1} = [\lambda] \tag{3.65}$$

The matrix on the right is a diagonal matrix containing eigenvalues as diagonal elements. All off-diagonal elements are zero. In this form matrix $[C]$ is recognized as the *diagonalization matrix*. We see that the diagonalization matrix is equivalent to the column matrix. Because this matrix is composed of eigenvectors, it is commonly called the "eigenvector matrix."

3.3.5   Calculating the Row Matrix

Rearranging (3.2) and then involving (3.64), we see that

$$[R] = [D][C]^{-1} = [D][C]^T \tag{3.66}$$

Having obtained matrix $[C]$ as described previously, we can calculate the complete set of numerical values for the elements of matrix $[R]$ by carrying out the multiplication shown in (3.66). Each element of the row matrix represents the "projection" of a row-designee point onto the respective eigenvector.

Before proceeding further, let us examine the row matrix in closer detail. The columns of this matrix are mutually orthogonal. Proof of this is given below, where we have made chronological use of (3.66), (3.17), (3.64), and (3.65), respectively:

$$
\begin{aligned}
[R]^T[R] &= ([D][C]^T)^T([D][C]^T) \\
&= [C][D]^T[D][C]^T \\
&= [C][Z][C]^{-1} \\
&= [\lambda]
\end{aligned}
\tag{3.67}
$$

From this result we conclude that

$$R'_j R_j = \lambda_j \tag{3.68}$$

and

$$R'_i R_j = 0 \tag{3.69}$$

where R_j, a column vector of the row matrix $[R]$, is associated with eigenvector

λ_j. Combining (3.31) and (3.68), we conclude that

$$R'_j R_j = \lambda_j = \sum_{i=1}^{r} r_{ij}^2 \tag{3.70}$$

This is in accord with (3.39) and (3.50), as expected.

From (3.70) we see that the eigenvalue is the sum of the squares of the projections of the row designees onto a given eigenvector. Since there are as many elements in R_j as there are rows in the data matrix, we see from (3.70) that each element of R_j represents the projection or score of that row designee onto the eigenvector axis C_j. For this reason the row matrix is generally called the projection matrix or score matrix.. Unlike C_j, R_j is not normalized, but can be normalized by dividing each element of R_j by the square root of the eigenvalue λ_j.

3.3.6 Short-Circuit Reproduction

Although there are c eigenvectors, only n principal eigenvectors are required to span the factor space. Consequently,

$$[D] \cong [D^{\ddagger}] = [R^{\ddagger}][C^{\ddagger}] = [R_1 \quad R_2 \quad \cdots \quad R_n] \begin{bmatrix} C'_1 \\ C'_2 \\ \vdots \\ C'_n \end{bmatrix} \tag{3.71}$$

There are n columns associated with the row matrix and n rows associated with the column matrix. The factor space is n-dimensional.

By following the procedures described in the earlier sections, we are able to find a row matrix and a column matrix which when multiplied together reproduce the data matrix within experimental error. Unfortunately, we cannot attach any physical significance to the resulting matrices since they represent mathematical solutions only. However, we can use these abstract results to classify row and column designees parametrically, and to reproduce the data empirically. Since the short-circuit reproduction scheme makes use of the mathematical, abstract factors of the space, we call this procedure abstract factor analysis (AFA).

In chemistry, measurements invariably contain experimental error. Errors that enter into the data matrix in a random fashion produce additional eigenvectors. Such eigenvectors have no real meaning. Their retention in the factor scheme unnecessarily increases the dimension of the factor space and yields predictions with an accuracy far beyond that which should be expected.

Because of experimental error, deciphering the dimension of the factor space is not an easy task. Various criteria have been developed for this purpose. These are discussed in considerable detail in Chapter 4. One simple criterion is based on comparing the original data matrix with that predicted from factor analysis using the short-circuit reproduction step. This method is described below.

To find the dimension of the factor space, we start with eigenvector C_1, associated with the largest eigenvalue, λ_1. This eigenvector is the most important and accounts for the maximum variance in the data. This fact is evident from (3.31), which shows that the eigenvalue is equal to the sum of the squares of the row-designee scores on the eigenvector axis; that is, $\lambda_j = \Sigma \, r_{ij}^2$, where the sum is taken over all the row elements. We now perform the following matrix multiplication:

$$[D]_1 = [R_1][C_1'] \tag{3.72}$$

where R_1 and C_1' are the respective vectors associated with λ_1. Matrix $[D]_1$ calculated in this way is compared to the original data matrix. If the agreement between the calculated and experimental matrices is not within experimental error, we continue the analysis by employing the next most important eigenvector:

$$[D]_2 = [R_1 \quad R_2] \begin{bmatrix} C_1' \\ C_2' \end{bmatrix} \tag{3.73}$$

If agreement is still not attained, we continue using the next important eigenvector and the next one, and so on, until we are able to reproduce the data satisfactorily. In this way we find that

$$[D]_n = [R_1 \quad R_2 \quad \cdots \quad R_n] \begin{bmatrix} C_1' \\ C_2' \\ \vdots \\ C_n' \end{bmatrix} = [D^{\ddagger}] \simeq [D] \tag{3.74}$$

where C_n is the last eigenvector axis needed to reproduce the data. The factor space is n-dimensional.

A word of caution is warranted at this time. If n is greater than either r or c (the number of rows or columns of the data matrix), either we have not introduced enough data (i.e., our data do not span the factor space) or all the data cannot be expressed in terms of the same factors (i.e., a large degree of uniqueness exists). On the other hand, if n is less than both r and c, this stage of the factor analysis is complete. It is good practice to strive for an r or c, whichever is smaller, to be at least twice n. This may be accomplished by increasing the number of columns or rows in the data matrix. The purpose of this precaution is to ensure that there are significantly more data points than unknown cofactors.

In this manner we have found a mathematical procedure for expressing the data as a product of two matrices in accord with (3.74). We have now completed the short-circuit reproduction loop shown in Figure 3.1 using the compressed factor space. This stage of the abstract factor analysis is complete.

3.4 TRANSFORMATION

From the standpoint of a theoretical chemist, the analysis should not terminate here. The row and column factors in their abstract forms are not recognizable as physical or chemical parameters, since the reference axes were generated to yield a purely mathematical solution. For scientific purposes we seek chemically recognizable factors. This can best be accomplished by transforming the reference axes so that they become aligned with fundamental properties of the designees.

3.4.1 Physically Significant Parameters

Within certain limitations we can transform the axes and find many solutions that obey (3.2). Consider, for example, points that lie in a common plane. The positions of such points can be designated by coordinates of two distinct axes that lie in the plane. These axes may be rotated freely in the plane and we may choose any two distinct axes to locate the points. In principle there are an infinite set of axes which can be used to define the plane and locate the data points. Similarly, in factor analysis we may rotate the reference axes as long as we keep them distinct and as long as they adequately span the space. In particular, we wish to transform the axes so that they are aligned with fundamental structural parameters of the row designees.

Transformation of the axes is accomplished by carrying out the following mathematical operation:

$$[\overline{R}] = [R^{\ddagger}][T] \tag{3.75}$$

where $[T]$ is the transformation matrix, of dimensions $n \times n$, and $[\overline{R}]$ is the row matrix in the new coordinate system. A least-squares method of obtaining the target transformation matrix is derived in Section 3.4.3.

The inverse of the transformation matrix is used to locate the column matrix in the new coordinate system. Equation (3.75) is first rearranged as follows:

$$[R^{\ddagger}] = [\overline{R}][T]^{-1} \tag{3.76}$$

This equation is placed into (3.2), giving

$$[D^{\ddagger}] = ([\overline{R}][T]^{-1})[C^{\ddagger}]$$
$$= [\overline{R}][T]^{-1}[C^{\ddagger}]$$
$$= [\overline{R}][\overline{C}] \tag{3.77}$$

where $[\overline{C}]$, the column matrix in the new coordinate system, is

$$[\overline{C}] = [T]^{-1}[C^{\ddagger}] \tag{3.78}$$

In summary, we have seen that upon proper transformation of the coordinate axes, it is possible to find a row matrix that can be interpreted in chemical or physical terms. Because there are an infinite number of positions through which a set of axes may be rotated, there exist an infinite number of possible solutions resulting from factor analysis. Nevertheless, only certain orientations of the axes yield factors that may correspond to recognizable parameters. It is advantageous to be able to test various physically significant parameters to determine whether or not they are the real factors. The least-squares method called target transformation is most appropriate for this purpose.

3.4.2   Techniques

In the classical work of factor analysis, the terminology "rotation" rather than "transformation" is most commonly encountered. One employs abstract rotations when there exists little or no information about the true origin of the factors. The ultimate goal of abstract rotation is to extract meaningful factors which have the simplest factor structure. All rotation techniques attempt to locate a set of axes so that as many row-designee points as possible lie close to the final factor axes, with only a small number of points remaining between the rotated axes.

There are many methods for obtaining rotation matrices. These methods are based upon some intuitive criteria of the factor space, such as simple structure, parsimony, factorial invariance, partialing, casual exploration, and hypothetical structure.[4] All these methods can be classified into one of two general categories: orthogonal rotations and oblique rotations. *Orthogonal rotations* preserve the angular relations between the original set of eigenvectors that emerge from FA. Techniques such as quartimax and varimax belong to this class. *Oblique rotations* do not preserve the angles between the eigenvectors. Oblimax, quartimin, biquartimin, covarimin, binormamin, maxplane, and promax fall in this class.

Target transformation belongs to the category of oblique rotations. Since this is by far the most important technique for chemists, we will study it in considerable detail. Those interested in learning details of the other methods listed above are advised to consult Rummel.[4] Before discussing target transformation, however, it is useful for us to examine briefly several of the popular methods of abstract rotations listed above.

Quartimax involves orthogonal rotation, preserving the angles between the eigenvector axes. The basic principle of quartimax is best visualized by a two-dimensional example. As the orthogonal axes are rotated so that one axis approaches a data point, the projection (i.e., the loading) of the point on the axis increases. At the same time the point moves away from the other axis and its loading on that axis decreases. Quartimax searches for a set of orthogonal axes that groups the points in clusters about each axis, with each point having either high or low loading on each axis. According to Harman,[5] this can be achieved by rotating the axes so as to maximize the quartimax function, Q:

$$Q = \sum_{j=1}^{n} \sum_{k=1}^{c} \lambda_j^2 \bar{c}_{jk}^4 \qquad (3.79)$$

Here λ_j is the jth eigenvalue and $\bar{c}_{jk}$ the loading of the kth data column vector on the jth axis after the rotation has been completed. The sum is taken over all c data columns and over n eigenvectors which are required to span the factor space. Unfortunately, quartimax tends to "overload" the first factor, producing one large general factor and many small subsidiary factors.

Varimax attempts to overcome the deficiency of the quartimax method. With varimax, the total variance, V, of the squared loadings is maximized:

$$V = \sum_{j=1}^{n} \left[\frac{1}{c} \sum_{k=1}^{c} (\lambda_j \bar{c}_{jk}^2)^2 - \frac{1}{c^2} \left(\sum_{k=1}^{c} \lambda_j \bar{c}_{jk}^2 \right)^2 \right] \qquad (3.80)$$

The varimax method of Kaiser[11] is currently the most popular of the orthogonal rotation schemes because of its ability to yield the same clusters regardless of the size of the data matrix.

Quartimin is essentially the same as quartimax except that the condition of orthogonality is removed. The eigenvector axes are rotated obliquely so that the loadings of a data point will be increased on one axis and decreased on all of the others. Thus the sum of the inner products of its loadings is reduced. According to Carroll,[12] this situation is achieved by minimizing the quartimin function, M:

$$M = \sum_{j<l=1}^{n} \sum_{k=1}^{c} \lambda_j \bar{c}_{jk}^2 \lambda_l \bar{c}_{lk}^2 \qquad (3.81)$$

Here j and l refer to the jth and lth oblique factors.

Oblimax involves oblique rotations where the number of low and high loadings on a given axis are increased by decreasing the loadings in the middle range. Saunders[13] showed that this can be accomplished by maximizing the kurtosis function, K,

$$K = \frac{\displaystyle\sum_{j=1}^{n} \sum_{k=1}^{c} \lambda_j^2 \bar{c}_{jk}^4}{\left(\displaystyle\sum_{j=1}^{n} \sum_{k=1}^{c} \lambda_j \bar{c}_{jk}^2 \right)^2} \qquad (3.82)$$

Covarimin[11] is an extension of the varimax method which permits oblique rotations. In this case the eigenvector axes are rotated obliquely until the covarimin function, C, is minimized:

$$C = \sum_{j<l=1}^{n} \left[\frac{1}{c} \sum_{k=1}^{c} \lambda_j \bar{c}_{jk}^2 \lambda_l \bar{c}_{lk}^2 - \frac{1}{c^2} \sum_{k=1}^{c} \lambda_j \bar{c}_{jk}^2 \lambda_l \bar{c}_{lk}^2 \right] \tag{3.83}$$

In general, covarimin usually yields axes that are similar to those resulting from varimax.

These methods have been extensively used in the behavioral sciences but have not been fully explored in chemistry.

3.4.3 Target Transformation

Because target transformation plays an important role in chemistry, we will discuss the underlying principles of this technique in detail. Target transformation is unique because, in spite of the complexity of the data space, it allows us to search for the basic factors *individually*. This can be seen by examining (3.75), which concerns the mathematical operation involved in transforming the eigenvector axes. From this equation we learn that $\overline{R}_l$, the lth column of the newly transformed row matrix, is obtained by multiplying T_l, the lth column of the transformation matrix, by the row matrix $[R^{\ddagger}]$:

$$\overline{R}_l = [R^{\ddagger}] T_l \tag{3.84}$$

We call $\overline{R}_l$ the *predicted vector* and T_l the associated *transformation vector*. We wish to find the transformation vector that yields an $\overline{R}_l$ most closely matching $\overline{\overline{R}}_l$, the *test vector* that we suspect is a basic factor. This test vector is our "target." To do this we carry out a least-squares procedure that minimizes the deviation between the test vector and the predicted vector. This procedure yields the best possible transformation vector for the individual target test being considered. The mathematical basis for obtaining the best T_l will now be described.

The transformation vector, T_l, has components $t_{1l}, t_{2l}, \ldots, t_{nl}$. Each row of $[R^{\ddagger}]$ can be looked upon as a row-designee vector. The ith row of $[R^{\ddagger}]$ is a vector R_i' having components $r_{i1}, r_{i2}, \ldots, r_{in}$. Vector R_i' should not be confused with R_i, the ith column of the row-factor matrix. When R_i' is multiplied, dot-product-wise, by T_l, we obtain $\bar{r}_{il}$, the projection of the ith row entity on the new target-transformed coordinate axis:

$$\bar{r}_{il} = R_i' \cdot T_l = r_{i1}t_{1l} + r_{i2}t_{2l} + \cdots + r_{in}t_{nl} \tag{3.85}$$

The sum in (3.85) is taken over all n principal factors.

Multiplying each row vector of the row matrix by T_l gives $\bar{r}_{1l}, \bar{r}_{2l}, \ldots, \bar{r}_{rl}$, which are the elements of $\overline{R}_l$, the predicted vector. Each element of the predicted vector is then compared to the corresponding element of the test vector $\overline{\overline{R}}_l$, having components $\overline{\overline{r}}_{1l}, \overline{\overline{r}}_{2l}, \ldots, \overline{\overline{r}}_{rl}$. The difference between the value of $\overline{r}_{il}$ and the value of $\overline{\overline{r}}_{il}$ is given as Δr_{il}:

$$\Delta r_{il} = \overline{r}_{il} - \overline{\overline{r}}_{il} = r_{il}t_{1l} + r_{i2}t_{2l} + \cdots + r_{in}t_{nl} - \overline{\overline{r}}_{il} \tag{3.86}$$

To find the best T_l, the deviation between the test vector and the predicted vector is minimized by setting the sum of the derivatives of the squares of the differences, given by (3.86), equal to zero. For example, the derivative of the square of the difference with respect to t_{11} is

$$\frac{d(\Delta r_{il})^2}{dt_{1l}} = 2r_{i1}^2 t_{1t} + 2r_{i1}r_{i2}t_{2l} + \cdots + 2r_{i1}r_{in}t_{nl} - 2r_{i1}\overline{\overline{r}}_{il} \tag{3.87}$$

Similar expressions are obtained for each row designee. Summing over all of the row designees and applying the least-squares criteria, we find that

$$\sum_{i=1}^{r} \frac{d(\Delta r_{il})^2}{dt_{1l}} = 0 = t_{1l} \sum_i r_{i1}^2 + t_{2l} \sum_i r_{i1}r_{i2} + \cdots$$

$$+ t_{nl} \sum_i r_{i1}r_{in} - \sum_i r_{i1}\overline{\overline{r}}_{il} \tag{3.88}$$

Repeating this calculation, we minimize the sum of squares of the deviation with respect to the remaining components of T_l: $t_{2l}, t_{3l}, \ldots, t_{nl}$. In this way we obtain the following set of simultaneous equations:

$$\sum r_{i1}\overline{\overline{r}}_{il} = t_{1l} \sum r_{i1}^2 \quad + t_{2l} \sum r_{i1}r_{i2} + \cdots + t_{nl} \sum r_{i1}r_{in}$$

$$\sum r_{i2}\overline{\overline{r}}_{il} = t_{1l} \sum r_{i1}r_{i2} + t_{2l} \sum r_{i2}^2 \quad + \cdots + t_{nl} \sum r_{i2}r_{in} \tag{3.89}$$

$$\vdots \qquad \vdots \qquad \vdots \qquad \qquad \vdots$$

$$\sum r_{in}\overline{\overline{r}}_{il} = t_{1l} \sum r_{il}r_{in} + t_{2l} \sum r_{i2}r_{in} + \cdots + t_{nl} \sum r_{in}^2$$

In order to express these equations in matrix form, we define the two vectors

$$A_l = \begin{bmatrix} \sum r_{i1}\overline{\overline{r}}_{il} \\ \sum r_{i2}\overline{\overline{r}}_{il} \\ \vdots \\ \sum r_{in}\overline{\overline{r}}_{il} \end{bmatrix} \quad \text{and} \quad T_l = \begin{bmatrix} t_{1l} \\ t_{2l} \\ \vdots \\ t_{nl} \end{bmatrix} \tag{3.90), (3.91}$$

and the following matrix:

$$[B] = \begin{bmatrix} \sum r_{i1}^2 & \sum r_{i1}r_{i2} & \cdots & \sum r_{i1}r_{in} \\ \sum r_{i1}r_{i2} & \sum r_{i2}^2 & \cdots & \sum r_{i2}r_{ij} \\ \vdots & \vdots & & \vdots \\ \sum r_{i1}r_{in} & \sum r_{i2}r_{in} & \cdots & \sum r_{in}^2 \end{bmatrix} \tag{3.92}$$

In matrix notation, the equations in (3.89) now become

$$A_l = [B]T_l \tag{3.93}$$

Multiplying both sides of this equation by $[B]^{-1}$, we obtain

$$T_l = [B]^{-1}A_l \tag{3.94}$$

Upon examining (3.92) and (3.67), we see that

$$[B] = [R^{\ddagger}]^T[R^{\ddagger}] = [\lambda^{\ddagger}] \tag{3.95}$$

where $[\lambda^{\ddagger}]$ is a diagonal matrix composed of the primary eigenvalues only. Furthermore, upon examining (3.90), we conclude that

$$A_l = [R^{\ddagger}]^T\overline{\overline{R}}_l \tag{3.96}$$

where $\overline{\overline{R}}_l$ is the test vector composed of the suspected parameters associated with the row designees. Thus

$$T_l = [\lambda^{\ddagger}]^{-1}[R^{\ddagger}]^T\overline{\overline{R}}_l \tag{3.97}$$

This equation is the central equation of target factor analysis. The least-squares vector transformer, T_l, a column of $[T]$, is readily calculated by means of this equation.

Because of (3.97), target transformation becomes a reality for the factor analyst. Numerical values for all quantities in this equation, except the test vector $\overline{\overline{R}}_l$, are automatically calculated during the routine decomposition step. The test vector, of course, must be obtained from theory, empirical knowledge, or intuition. Deducing the test vector constitutes the real chemical art involved in target factor analysis. To see whether or not a suspected factor is a true factor, one inserts the chosen test vector into (3.97) and the best possible transformation vector emerges.

Having obtained T_l, we use (3.84) to obtain numerical values for the elements of $\overline{R}_l$. We can then ascertain whether or not the following equation is obeyed within experimental error:

$$\overline{R}_l \overset{?}{=} \overline{\overline{R}}_l \tag{3.98}$$

If our suspected test vector $\overline{\overline{R}}_l$ is a factor, each element of $\overline{R}_l$ will equal the corresponding element of $\overline{\overline{R}}_l$, within experimental error. If, on the contrary, it is not a true factor, the differences between the corresponding elements of $\overline{R}_l$ and $\overline{\overline{R}}_l$ will be greater than our expectations. Methods for judging the validity of a test vector are described in Section 4.6.

3.4.4 Free Floating

The least-squares method for transformation as developed in the previous section is completely general. It holds even if some of the $\bar{\bar{r}}_{il}$ values for a particular test vector are missing. In this case, however, appropriate terms must be removed from the summations in (3.89) through (3.94). Vector T_l must then be calculated by means of (3.94) since (3.97) would no longer be valid. With these appropriate modifications we can calculate $\bar{R}_l$ by means of (3.84), and compare the results with the experimental test vector, $\bar{\bar{R}}_l$. This procedure, termed *free floating*, has a hidden advantage. Equation (3.84) automatically yields an $\bar{r}_{il}$ value for each designee, including those free-floated in the test vector.

There is one important restriction on the target test. The number of test points in a test vector must be at least equal to n, the rank of the data matrix. The introduction of a test vector with insufficient test points will always yield a perfect fit of the test points. Such results, however, are totally meaningless. When using the target test procedure, extreme care must be exercised to include more test points than factors. For a detailed discussion of this point, refer to Section 4.6.

To illustrate free floating we again return to Figure 3.4, which shows the vector relationship between four columns (four properties) of a given data matrix. Using the least-squares target transformation technique, we attempt to find a reference axis that has physical significance and lies in this plane. If the dipole moment were a true factor, the projection of a row-designee (molecule) point on the dipole axis would yield the dipole moment of the molecule. To locate this axis, we need not know the dipole moments of all row-designee molecules. For a two-factor space a minimum of two dipole moments is required mathematically. However, it is advisable to use more than two test points.

In practice, a dipole moment test vector is constructed by free-floating all row-designee molecules whose dipole moments are not known. The transformation vector is calculated using only the partial set of data points, as described earlier. The computer output, which is the result of (3.84), yields dipole moments for all molecules, whether or not they were included in the test scheme. In this manner, dipole moments can be predicted. Although this was not the original intent of factor analysis, it does constitute an extremely useful and important fringe benefit.

3.4.5 Combination

It is possible to find a sufficient number of acceptable test vectors, but still not span the factor space, because some of the test vectors lie in a common subspace. To ascertain whether or not a particular set of test vectors adequately spans the space, we perform the following calculation:

$$[\overline{\overline{R}}][\overline{C}] = [D^{\ddagger}]_{\text{TFA}} \overset{?}{=} [D] \qquad (3.99)$$

Each column of $[\overline{\overline{R}}]$ is an individually successful test vector that satisfactorily obeys (3.98). Matrix $[\overline{C}]$ is obtained by means of (3.78). $[D^{\ddagger}]_{\text{TFA}}$ is the reproduced data matrix that results from this target-combination test. If this matrix matches $[D]$, the original data matrix, within experimental error, we know that a proper set of real factors has been found.

In (3.99), we use row matrix $[\overline{\overline{R}}]$ instead of $[\overline{R}]$. If, instead of $[\overline{\overline{R}}]$ we inadvertently use $[\overline{R}]$, we would simply perform a unitary transformation, obtaining exactly the same data matrix as was originally obtained from AFA, with the same number of factors. In other words,

$$[\overline{R}][\overline{C}] = [R^{\ddagger}][T][T]^{-1}[C^{\ddagger}] = [R^{\ddagger}][C^{\ddagger}] = [D^{\ddagger}] \cong [D] \qquad (3.100)$$

In effect, we would be forcing the model to fit the data. On the other hand, (3.99) is *not* a unitary transformation and does not force the model to fit the data. Consequently, (3.99) is a unique and severe test of the model. This important distinction between (3.99) and (3.100) cannot be overstressed.

3.4.6   Key Combination Sets

Basic Factors. The ultimate objective of factor analysis is to obtain a basic set of factors that have real physicochemical meaning (see Section 2.4). Basic vectors can be tested individually and identified by applying (3.97), (3.84), and (3.98) as described in Section 3.4.5. Because of the variety of ways of expressing real chemical factors and because of the multifaceted interrelationships among the factors, finding a "key set" of basic factors that adequately spans the data space is not a simple task. To find the key set, various combinations of acceptable basic vectors are formed into row-factor matrices, $[\overline{\overline{R}}]_{\text{basic}}$, and the following calculation is carried out:

$$[\overline{\overline{R}}]_{\text{basic}}[\overline{C}]_{\text{basic}} = [D^{\ddagger}]_{\text{basic}} \qquad (3.101)$$

where

$$[\overline{C}]_{\text{basic}} = [T]^{-1}[C^{\ddagger}] \qquad (3.102)$$

Here $[\overline{C}]_{\text{basic}}$, the basic column-factor matrix, is calculated by premultiplying the abstract column-factor matrix by the inverse of the transformation matrix constructed from the individual transformation vectors associated with the basic vectors. A key set of basic vectors is found when $[D^{\ddagger}]_{\text{basic}}$, the combination reproduced data, adequately equals the original data matrix:

$$[D^{\ddagger}]_{\text{basic}} \simeq [D] \qquad (3.103)$$

Typical Factors. Since the columns of the data matrix lie in the factor space, a judicious choice of n data columns can be used to describe the n-dimensional factor space. Such a combination is called a key set of typical vectors.

To understand this, we return to Figure 3.3, which illustrates the vector relationship between four columns (four data vectors) of a given data matrix. This figure shows us that the four data vectors lie in a common plane defined by two eigenvectors which emerge from the factor analytical decomposition step. Since all four data vectors lie in the plane, any two data-column vectors can be used to locate a row-designee point. Predictions can then be made by reading the projections on the remaining data-column vectors. By means of the least-squares target transformation approach, the reference axes can be reassigned to coincide with an appropriate set of typical columns of the original data matrix. This is accomplished by using a column of the original data as a test vector. Obviously, each column will yield a successful test, since each column lies in the factor space. The test procedure, however, will produce a transformation vector. A combination of such transformation vectors is needed to construct the transformation matrix. In general, an arbitrary combination of data columns will not necessarily yield a transformation matrix that will be successful in reproducing the data. Only certain combinations, which span the total factor space, will work. The combination set that best accomplishes this task is called the *key combination set*.

The mathematical steps involved in this process are as follows. The lth column of the data matrix can be looked upon as a vector $\overline{\overline{D}}_l$ with the data points in the column representing the elements of the vector. In other words, $\overline{\overline{D}}_l$ is used as a test vector. Letting $\overline{\overline{D}}_l = \overline{\overline{R}}_l$ and using (3.97), we find that

$$T_l = [\lambda]^{-1}[R]^T\overline{\overline{D}}_l \qquad (3.104)$$

Using n of these transformation vectors in combination, we construct a complete transformation matrix:

$$[T] = [T_a \quad T_b \quad \cdots \quad T_n] \qquad (3.105)$$

Various combinations are employed in an attempt to find the key combination set that best reproduces the data matrix. Hence we search for a set of n data columns that best satisfies

$$[\overline{\overline{D}}]_{key}[\overline{C}] - [D] = minimum \qquad (3.106)$$

where

$$[\overline{\overline{D}}]_{key} = [\overline{\overline{D}}_a \quad \overline{\overline{D}}_b \quad \cdots \quad \overline{\overline{D}}_n] \qquad (3.107)$$

and

$$[\overline{C}] = [T]^{-1}_{key}[C^{\ddagger}] \qquad (3.108)$$

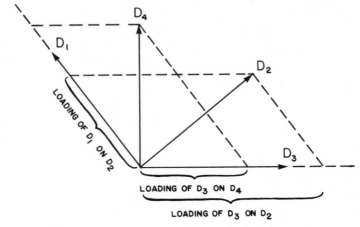

Fig. 3.5. Factor loadings onto representative axes D_1 and D_3. Lines drawn parallel to D_1 and D_3 intersect D_1 and D_3 at points that correspond to the loadings.

Equation (3.106), involving typical vectors, is analogous to (3.102), involving basic vectors. In these expressions we insert the subscript "key" to signify the key combination set.

There are several advantages to using data columns as factors of the space. First, they are more easily visualized than the abstract factors produced by factor analysis. Second, their use precludes the need to identify the true controlling factors. Third, empirical predictions can be made quickly from the resulting equations. For many chemical problems, this is sufficient. Powerful uses of this technique are found in later chapters.

To further illustrate the principle involved in using the key combination set, let us return to our example portrayed in Figures 3.2 and 3.3. Here we are dealing with four data column vectors that lie in a common plane. Because all four data column vectors lie in the same plane, they are linearly dependent. Any two of the four vectors shown in Figure 3.2 can be used as the representative axes. For example, we may choose D_1 and D_3 to be the bases. From the tips of vectors D_2 and D_4 we draw lines parallel to the basic set D_1 and D_3, as shown in Figure 3.5. The lengths of the intersections on each basis axis represents the loadings of the basic vectors on the vectors in question. From this diagram we can show that the four vectors can be represented as follows:

$$D_1 = 1.0000D_1 + 0.0000D_3$$
$$D_2 = 0.7945D_1 + 1.2330D_3$$
$$D_3 = 0.0000D_1 + 1.0000D_3 \qquad (3.109)$$
$$D_4 = 1.2361D_1 + 0.7265D_3$$

These linear equations have an important feature. The projection of a row-designee point on D_2 and D_4 can be calculated from the scores on D_1 and D_3, the basis axes. Thus a data set consisting of a large array of data columns can be reduced to a minimum set of representative data columns from which all the data can be generated. With the use of these equations, we no longer need to compute the projections on the abstract eigenvectors resulting from factor analysis. Instead, we need only to store the data involving the chosen representative axes, D_1 and D_3.

3.4.7 Column Matrix Transformation Test

Instead of focusing our attention on row matrix transformations, we could have focused our attention on column matrix transformations. When we have sufficient information concerning row factors, we search for row transformation vectors. Conversely, when we have information concerning column factors, we search for column transformation vectors instead. Although we focused attention in previous sections solely on testing row factors, we could easily develop analogous equations for testing column factors. Since the mathematical development is quite similar to that given previously, we will not repeat the details here.

When we wish to transform the column matrix into our column factor model matrix, we employ the equation

$$[T][C^{\ddagger}] = [\overline{C}] \tag{3.110}$$

Column transformation will yield an expression analogous to (3.99):

$$[\overline{R}][\overline{\overline{C}}] = [D^{\ddagger}]_{\text{TFA}} \overset{?}{=} [D] \tag{3.111}$$

In (3.111), $[\overline{\overline{C}}]$ is the test factor matrix and $[\overline{R}]$ is the loading matrix, whereas in (3.99), $[\overline{\overline{R}}]$ is the test factor matrix and $[\overline{C}]$ is the loading matrix.

REFERENCES

1. E. R. Malinowski, *Diss. Abstr.*, **23** (8), Publ. No. 62-2027 (1963).
2. P. T. Funke, E. R. Malinowski, D. E. Martire, and L. Z. Pollara, *Sep. Sci.*, **1**, 661 (1966).
3. P. H. Weiner, E. R. Malinowski, and A. R. Levinstone, *J. Phys. Chem.*, **74**, 4537 (1970).
4. R. J. Rummel, *Applied Factor Analysis*, Northwestern University Press, Evanston, Ill., 1970.
5. H. H. Harman, *Modern Factor Analysis*, 3rd ed. rev., University of Chicago Press, Chicago, 1967.
6. B. Fruchter, *Introduction to Factor Analysis*, D. Van Nostrand, Princeton, N.J., 1954.

7. A. L. Comrey, *A First Course in Factor Analysis,* Academic Press, New York, 1973.
8. D. N. Lawley and A. E. Maxwell, *Factor Analysis as a Statistical Method,* 2nd ed., Butter-worths, London, 1971.
9. P. Horst, *Factor Analysis of Data Matrices,* Holt, Rinehart and Winston, New York, 1965.
10. R. W. Rozett and E. M. Petersen, *Anal. Chem.,* **47,** 1301 (1975).
11. H. F. Kaiser, *Psychometrika,* **23,** 187 (1958).
12. J. B. Carroll, *Psychometrika,* **18,** 23 (1953).
13. D. R. Saunders, *Psychometrika,* **25,** 199 (1960).

4 Effects of Experimental Error

4.1 THEORY OF ERROR

In the absence of experimental error, factor analysis will yield the exact number of factors involved in a data matrix. Unfortunately, perfect data are not attainable, and the chemist is plagued with handling data that possess experimental uncertainty. Although in many instances we can make a reliable estimate of the error, we are often not sure of the uncertainty estimation. Experimental error blends into FA and tends to complicate the process at every decision-making step. For this reason it is important for us to study how errors weave into the factor analytical scheme. Only then can we make provisions to account for the perturbations produced by error. We learn in this chapter how special factor analytical techniques can be used to deduce not only the number of factors but also the experimental error, even when no such estimates are available.

4.1.1   Preliminary Considerations

Because of experimental error a "raw" data point, d_{ik}, is best represented as a sum of two terms:

$$d_{ik} = d^*_{ik} + e_{ik} \tag{4.1}$$

where d^*_{ik} represents a "pure" data point, free of experimental error, and e_{ik} is the experimental error associated with the data point.

Experimental error invariably produces a larger number of eigenvectors than is required by the pure factor space. Retention of all the resulting eigenvectors will lead to perfect reproduction of the raw data, including experimental error. Hence we need reliable criteria to help us choose the correct number of eigenvectors. Even when we do select the correct eigenvectors, it is impossible to remove all the error. In fact, we shall find that a part of the error mixes into the data reproduction scheme. In other words,

$$e_{ik} = e^\dagger_{ik} + e^0_{ik} \tag{4.2}$$

where $e^\dagger_{ik}$ is that part of the error which mixes into the factor analytical scheme and e^0_{ik}, called the *residual error*, is that part which can be removed by deleting the unnecessary eigenvectors.

Inserting (4.2) into (4.1), we see that

$$d_{ik} = d^\dagger_{ik} + e^0_{ik} \tag{4.3}$$

where

$$d^\dagger_{ik} = d^*_{ik} + e^\dagger_{ik} \tag{4.4}$$

Here $d^\dagger_{ik}$ is a reproduced data point using the proper number of eigenvectors. From (4.4) we see that the reproduced data point contains some error. From (4.3) we see that the residual error e^0_{ik} is simply the difference between the raw data and reproduced data point. Criteria for choosing the correct number of eigenvectors must be based upon an understanding of how the experimental error enters and mixes into the factor analytical scheme. In this section we investigate this important aspect.

The raw data matrix is simply the sum of two matrices: a pure data matrix $[D^*]$ and an error matrix $[E]$:

$$[D] = [D^*] + [E] \tag{4.5}$$

All these matrices have the same size, $r \times c$, involving r rows and c columns. However, their dimensionalities are not the same. Although the pure data matrix is n-dimensional, the raw data matrix and the error matrix are either r- or c-dimensional, whichever is smaller. Throughout our discussion here we assume that c is smaller than r. Hence the error matrix is c-dimensional.

Since the factor space of the pure data is n-dimensional, we can express a pure data point as a linear sum of n product terms:

$$d_{ik}^* = \sum_{j=1}^{n} r_{ij}^* c_{jk}^* \qquad (4.6)$$

This expression is identical to (3.1), except that we have inserted an asterisk as a superscript to distinguish the pure factors from factors that contain an amalgamation with error.

Although the error matrix is the same size as the data matrix, a larger number of eigenvector axes are required to span the error space than the pure data space. This is due to the fact that the error matrix consists of random values. Any orthogonal set of axes can be chosen to define the error space, as long as there is a sufficient number of axes. The number of axes required to span the error space will exactly equal the number of columns in the data matrix, assuming that $c < r$.

A set of n basis axes, called *primary axes,* is required to describe the raw data space within experimental error. However, all c axes are required to span the error space. In other words, there is a set of axes ($c - n$ in number), called *error axes* or *secondary axes,* which are associated solely with the remaining part of the error. Since the same n axes used to describe the raw data space may be used to describe part of the error space, the error associated with the raw data point, d_{ik}, can be represented as the following linear sum:

$$e_{ik} = \sum_{j=1}^{n} \sigma_{ij}^\dagger c_{jk} + \sum_{j=n+1}^{c} \sigma_{ij}^0 c_{jk} \qquad (4.7)$$

Here c_{jk} is the kth component of the jth principal axis, σ_{ij} the projection of the ith row designee of the error matrix onto the jth primary axis, and σ_{ij}^0 the corresponding projection onto the jth secondary axis. The sum is taken over all c eigenvector axes.

Placing (4.6) and (4.7) into (4.1), we find that

$$d_{ik} = \sum_{j=1}^{n} (r_{ij}^* c_{jk}^* + \sigma_{ij}^\dagger c_{jk}) + \sum_{j=n+1}^{c} \sigma_{ij}^0 c_{jk} \qquad (4.8)$$

Defining r_{ij} as follows:

$$r_{ij} = r_{ij}^* \frac{c_{jk}^*}{c_{jk}} + \sigma_{ij}^\dagger \qquad (4.9)$$

we obtain

$$d_{ik} = \sum_{j=1}^{n} r_{ij}c_{jk} + \sum_{j=n+1}^{c} \sigma_{ij}^{0}c_{jk} \tag{4.10}$$

Based upon this formulation, the complete factor analytical solution in matrix notation can be expressed as

$$[D] = [R^{\#}][C] + [R^{0}][C] \tag{4.11}$$

Here $[C]$ is composed of the complete set of eigenvectors,

$$[R^{\#}] \equiv \begin{bmatrix} r_{11} & \cdots & r_{1n} & 0 & \cdots & 0 \\ r_{21} & \cdots & r_{2n} & 0 & \cdots & 0 \\ \vdots & & \vdots & \vdots & & \vdots \\ r_{r1} & \cdots & r_{rn} & 0 & \cdots & 0 \end{bmatrix} \tag{4.12}$$

and

$$[R^{0}] \equiv \begin{bmatrix} 0 & \cdots & 0 & \sigma_{1,n+1}^{0} & \cdots & \sigma_{1c}^{0} \\ 0 & \cdots & 0 & \sigma_{2,n+1}^{0} & \cdots & \sigma_{2c}^{0} \\ \vdots & & \vdots & \vdots & & \vdots \\ 0 & \cdots & 0 & \sigma_{r,n+1}^{0} & \cdots & \sigma_{rc}^{0} \end{bmatrix} \tag{4.13}$$

If the zero elements, which are associated with the secondary eigenvalues, of $[R^{\#}]$ are dropped from this $r \times c$ matrix, the usual $r \times n$ row factor matrix, $[R^{\ddagger}]$, is obtained. Similarly, upon deleting the secondary eigenvectors, $[C]$ is reduced to $[C^{\ddagger}]$, the usual $n \times c$ column-factor matrix. The elements of $[R^{0}]$ consist of nothing but error components and contain no useful information. Deletion of this matrix leads to factor compression, and a reproduced data matrix, $[D]$, which differs slightly from the raw data matrix but is essentially the same within experimental error:

$$[D^{\ddagger}] = [R^{\#}][C]$$

$$= [R^{\ddagger}][C^{\ddagger}] \tag{4.14}$$

where $[C^{\ddagger}]$ contains only the n primary eigenvectors.

By following the reasoning given in Section 3.3.1, we find that the covariance matrix can be decomposed into the following sum:

$$[Z] = \sum_{j=1}^{n} \lambda_{j}^{\ddagger} C_{j} C_{j}' + \sum_{j=n+1}^{c} \lambda_{j}^{0} C_{j} C_{j}' \tag{4.15}$$

where

$$\lambda_{j}^{\ddagger} = \sum_{i=1}^{r} r_{ij}^{2} \qquad \text{for} \quad j = 1, \ldots, n \tag{4.16}$$

and

$$\lambda_j^0 = \sum_{i=1}^{r} (\sigma_{ij}^0)^2 \qquad \text{for} \quad j = n + 1, \ldots, c \qquad (4.17)$$

These equations show exactly how the error mixes into the factor analytical scheme. If there were no error, $\sigma_{ij}^\dagger$ and σ_{ij}^0 would equal zero, c_{jk} would equal c_{jk}^*, and (4.16) would reduce to

$$\lambda_j^\dagger = \sum_{i=1}^{r} r_{ij}^{*2} \qquad (4.18)$$

and all the λ_j^0 would vanish. Instead of c eigenvalues, only n eigenvalues would result. The factor space would be clearly identified.

To see how the error mixes into the eigenvectors, we take advantage of the symmetry of the problem. The error could have been expressed in terms of axes associated with r_{ij} rather than c_{jk}. Instead of (4.7), we can write

$$e_{ik} = \sum_{j=1}^{n} r_{ij}\sigma_{jk}^\dagger + \sum_{j=n+1}^{c} r_{ij}\sigma_{jk}^0 \qquad (4.19)$$

where r_{ij} is the ith component of the jth axis, and $\sigma_{jk}^\dagger$ and σ_{jk}^0 are the projections of the kth column designee of the error matrix onto the jth axis of the primary and secondary axes, respectively.

We now make the following deductions, analogous to those employed to obtain (4.8):

$$d_{ik} = \sum_{j=1}^{n} (r_{ij}^* c_{jk}^* + r_{ij}\sigma_{jk}^\dagger) + \sum_{j=n+1}^{c} r_{ij}\sigma_{jk}^0$$

$$= \sum_{j=1}^{n} r_{ij}c_{jk} + \sum_{j=n+1}^{c} r_{ij}\sigma_{jk}^0 \qquad (4.20)$$

where c_{jk} is defined as

$$c_{jk} \equiv c_{jk}^* \frac{r_{ij}^*}{r_{ij}} + \sigma_{jk}^\dagger \qquad (4.21)$$

This equation shows precisely how the experimental error perturbs the components of the eigenvectors of factor analysis. If there were no error, σ_{jk}^0 and $\sigma_{jk}^\dagger$ would be zero, $r_{ij}^\dagger$ would equal r_{ij}^*, and c_{jk} would equal c_{jk}^*, as expected.

4.1.2   Primary and Secondary Factors

If the data matrix contained no error, the covariance matrix would be decomposed into a sum of n factors.

$$[Z] = \sum_{j=1}^{n} \lambda_j C_j C_j' \qquad (4.22)$$

However, because of experimental error, the decomposition will lead to a larger number of eigenvectors. In fact, since the covariance matrix is of size $c \times c$, where c is the number of columns in the data matrix, the decomposition will yield c factors. That is,

$$[Z] = \sum_{j=1}^{c} \lambda_j C_j C_j' \tag{4.23}$$

As shown in (4.15), this sum can be separated into two groups. The first n terms in this sum are associated with the true factors but contain an admixture of error. The second set of terms consists of pure error. It is this second set of terms that should be omitted from further consideration.

The trace of covariance matrix is invariant upon the similarity transformation expressed by (3.65). Hence the trace of the covariance matrix is related to the data points and the eigenvalues as follows:

$$\sum_{i=1}^{r} \sum_{k=1}^{c} d_{ik}^2 = \text{trace } [Z] = \sum_{j=1}^{c} \lambda_j \tag{4.24}$$

Because of experimental error, the number of eigenvalues will always equal the number of columns in the data matrix, assuming, of course, that $c < r$. These eigenvalues can be grouped into two sets: a set consisting of the true eigenvalues but containing an admixture of error, and a set consisting of error only:

$$\sum_{j=1}^{c} \lambda_j = \sum_{j=1}^{n} \lambda_j^{\ddagger} + \sum_{j=n+1}^{c} \lambda_j^0 \tag{4.25}$$

The first set, called the *primary eigenvalues,* consists of the first n members having the largest values. The second set, called the *secondary eigenvalues,* contains $(c - n)$ members having the smallest values. The secondary eigenvalues are composed solely of experimental error, and their removal from further consideration will, in fact, lead to data reproduction that will be more accurate than the original data (see Section 4.2). We must develop criteria to deduce how many of the smallest eigenvalues belong to this set.

Because the primary set of eigenvalues contain an admixture of error, their associated eigenvectors are not the true eigenvectors. Factor analysis of the improved data matrix, regenerated by the primary set of eigenvectors, will yield exactly the same set of primary eigenvalues and eigenvectors. The new secondary set will be extremely small, because it is due to numerical roundoff. Consequently, further purification of the data by repeated factor analytical decomposition is not possible.

Using the primary eigenvectors, we can regenerate the data, which we label $d_{ik}^{\dagger}$. The regenerated data obey the following relationship analogous to (4.24):

$$\sum_{i=1}^{r} \sum_{k=1}^{c} d_{ik}^{\ddagger 2} = \text{trace } [Z^{\ddagger}] = \sum_{j=1}^{n} \lambda_j^{\ddagger} \tag{4.26}$$

By subtracting (4.26) from (4.24), we find that

$$\sum_{i=1}^{r} \sum_{k=1}^{c} (d_{ik}^2 - d_{ik}^{\ddagger 2}) = \sum_{j=n+1}^{c} \lambda_j^0$$

$$= \sum_{i=1}^{r} \sum_{j=n+1}^{c} (\sigma_{ij}^0)^2 \tag{4.27}$$

where σ_{ij}^0 is a row cofactor associated with the residual error. In other words, the sum of the differences between the squares of the raw data and the reproduced data equals the sum of the secondary eigenvalues. This sum is associated with the error removed by neglecting the secondary eigenvalues. Equation (4.27) is important because it shows the relationship among the raw data, the reproduced data, and the secondary eigenvalues.

Although the pure data matrix and error matrix are not mutually orthogonal, it is quite surprising that the reproduced data matrix and its associated residual error matrix are mutually orthogonal. To prove this, we first consider the perfect reproduction of the data matrix using all the eigenvectors, as shown in (4.11). This equation reproduces the data perfectly, including all the error. From (4.11) and (4.14), we conclude that

$$[D] = [D^{\ddagger}] + [E^0] \tag{4.28}$$

where

$$[E^0] = [R^0][C]$$

We note here that the residual error matrix, $[E^0]$, is the difference between the raw data matrix and the AFA-reproduced data matrix. Furthermore, from (4.12) and (4.13), it is evident that

$$[R^{\#}]^T [R^0] = [0] \tag{4.29}$$

Using this equation, we find that

$$\begin{aligned}
[D^{\ddagger}]^T [E^0] &= \{[R^{\#}][C]\}^T \{[R^0][C]\} \\
&= [C]^T [R^{\#}]^T [R^0][C] \\
&= [C]^T [0][C] \\
&= [0] \tag{4.30}
\end{aligned}$$

Thus we conclude that the AFA-regenerated data matrix and its associated residual error matrix are mutually orthogonal. We will use this important fact later to develop error criteria for determining the dimensions of the factor space.

4.1.3 Example

To illustrate how the experimental error perturbs the primary eigenvectors and produces secondary eigenvectors (which are composed solely of error), let us examine the simple one-factor data matrix[1,2] shown in Table 4.1. This matrix consists of two identical data columns labeled "pure data matrix." This matrix is obviously one-dimensional since a plot of the points of the first column against the corresponding points of the second column yields a perfectly straight line with all points lying exactly on the one-dimensional line axis. When this data matrix is factor-analyzed, via the covariance matrix, the results shown in Table 4.2 are obtained. Each of the row cofactors, r_{i1}^*, is the distance from the origin to the data point. The eigenvalue, 770, is simply the sum of the squares of the row cofactors (the scores).

To see how experimental error perturbs these results, the following procedure is employed. First the "error matrix" shown in Table 4.1 is arbitrarily generated. This error matrix is then added to the pure data matrix to give the "raw data matrix," also shown in Table 4.1. The raw data matrix simulates real chemical data possessing experimental error. When the raw data matrix is factor-analyzed, not one but two eigenvalues and their associated eigenvectors are produced. These are listed in Table 4.2.

Table 4.1 **Values used to construct an artificial raw data matrix and the results of factor-analyzing this raw data matrix**[a]

Pure Data Matrix, $[D^*]$		Error Matrix, $[E]$		Raw Data Matrix, $[D] = [D^*] + [E]$		Reproduced Data Matrix Using One Factor, $[D^{\ddagger}] = [R][C]$	
1	1	0.2	0.0	1.2	1.0	1.0936	1.1052
2	2	−0.2	−0.2	1.8	1.8	1.7904	1.8095
3	3	−0.1	0.1	2.9	3.1	2.9846	3.0163
4	4	0.0	−0.1	4.0	3.9	3.9288	3.9705
5	5	−0.1	0.0	4.9	5.0	4.9240	4.9763
6	6	0.2	−0.2	6.2	5.8	5.9671	6.0305
7	7	0.2	−0.1	7.2	6.9	7.0118	7.0862
8	8	−0.2	0.1	7.8	8.1	7.9086	7.9926
9	9	−0.2	0.1	8.8	9.1	8.9033	8.9978
10	10	−0.1	0.2	9.9	10.2	9.9974	10.1036

[a] Reprinted with permission from E. R. Malinowski, *Anal. Chem.*, **49**, 606 (1977).

Table 4.2   **Eigenvalues and row cofactors resulting from factor analysis**[a]

From Pure Data Matrix	From Raw Data Matrix	
$\lambda_1 = 770$	$\lambda_1^{\dagger} = 767.1514$	$\lambda_2^0 = 0.2886124$
r_{i1}^{*}	r_{i1}	σ_{i2}^0
1.41421	1.55487	0.14964
2.82843	2.54555	0.01344
4.24264	4.24333	−0.11901
5.65685	5.58569	0.10021
7.07107	7.00063	−0.03374
8.48528	8.48367	0.32765
9.89950	9.96895	0.26479
11.31371	11.24396	−0.15275
12.72792	12.65816	−0.14528
14.14214	14.21377	−0.13707

[a] Reprinted with permission from E. R. Malinowski, *Anal. Chem.*, **49**, 606 (1977).

Two eigenvectors emerge because the raw data points, unlike the pure data points, lie in a two-dimensional plane and not on a one-dimensional line. This is dramatically illustrated in Figure 4.1, where the points of the first column of

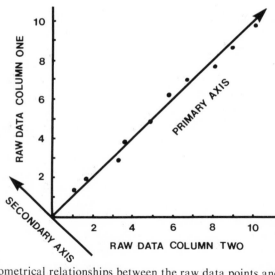

Fig. 4.1. Geometrical relationships between the raw data points and the primary and secondary axes resulting from principal factor analysis of the raw data matrix [Reprinted with permission from E. R. Malinowski, *Anal. Chem.*, **49**, 606 (1977)].

the raw data matrix are plotted against the corresponding points of the second column. The direction of the primary axis is shifted slightly from its original angle of 45° so that $C_1 \neq C_1^*$. This occurs because part of the error cannot be removed by deleting the $c - n$ secondary eigenvectors. Each row cofactor, r_{i1}, is the distance along the primary axis to the point where the perpendicular projection of the data point intersects the axis. Notice that the primary row cofactors of the raw data matrix are different from those of the pure data matrix. One can verify, after some tedious calculations, that these values obey (4.9).

The second eigenvector, C_2, which emerges consists of pure error. Each row cofactor, σ_{i2}^0, is the distance along the secondary axis (see Figure 4.1) from the origin to the place where the projection of the raw data point intersects the secondary axis. The secondary eigenvalue, λ_2^0, has no physical meaning, since it is the sum of squares of the secondary row cofactors, which contain nothing but error. By deleting this secondary axis one obtains a reproduced raw data matrix that emulates the pure data matrix more accurately than does the original raw data matrix. This can be verified by comparing the reproduced and the raw data matrices with the pure matrix. The reproduced data matrix, using only the primary eigenvector, is given on the extreme right in Table 4.1.

4.2   AFA FOR DATA IMPROVEMENT

AFA short-circuit reproduction has a built-in statistical feature. Because the secondary set of abstract eigenvalues are composed of pure error, their removal invariably will lead to data improvement. Like magic, this is accomplished without any a priori knowledge of the controlling factors. Although this is not the original intent of factor analysis, it does represent a valuable, unexpected benefit. In this section we will attempt to determine quantitatively how much data improvement is possible using abstract factor analysis.

The raw experimental data matrix is, in reality, a sum of two matrices: a pure data matrix, having no error, and an error matrix. Our discussion in Section 4.1.2 shows that when we factor-analyze the *pure* data matrix we obtain exactly n eigenvalues and n eigenvectors. If, on the other hand, we factor-analyze the *raw* data, we obtain c eigenvalues and c eigenvectors. However, only n of these are associated with the true factors. The remaining $(c - n)$ eigenvalues and eigenvectors are composed of pure error.

Instead of factor-analyzing the raw data matrix, let us examine what happens when we factor-analyze the error matrix via its covariance matrix. The same eigenvectors used to describe the raw data can be used to describe the error space.[1,2] Of course, the resulting eigenvalues would be composed of pure error. Instead of (4.25), we would obtain the following analogous equation:

$$\sum_{j=1}^{c} \lambda_{je} = \sum_{j=1}^{n} \lambda_{je}^{\dagger} + \sum_{j=n+1}^{c} \lambda_{je}^{0} \tag{4.31}$$

where

$$\sum_{j=1}^{n} \lambda_{je}^{\dagger} = \sum_{i=1}^{r} \sum_{j=1}^{n} (\sigma_{ij}^{\dagger})^2 \tag{4.32}$$

and

$$\sum_{j=n+1}^{c} \lambda_{je}^{0} = \sum_{i=1}^{r} \sum_{j=n+1}^{c} (\sigma_{ij}^{0})^2 \tag{4.33}$$

In (4.31), the term on the left is equal to the sum of the squares of all the error points in the data matrix. This is true because the trace of the covariance matrix constructed from the error matrix is invariant upon the similarity transformation involved in the decomposition (diagonalization) process. Combining this fact with (4.31), (4.32), and (4.33), we find that

$$\sum_{i=1}^{r} \sum_{k=1}^{c} e_{ik}^2 = \sum_{i=1}^{r} \sum_{j=1}^{n} (\sigma_{ij}^{\dagger})^2 + \sum_{i=1}^{r} \sum_{j=n+1}^{c} (\sigma_{ij}^{0})^2 \tag{4.34}$$

The term on the left is the sum of the experimental error squared. It also represents the sum of the squares of the projections of the error points onto all c data column axes. The first sum on the right concerns the projections of the error points onto the n primary eigenvector axes. This sum represents the error that mixes into the factor analytical process and cannot be removed. The second sum on the right concerns the projections onto the $(c - n)$ secondary axes, the axes that are removed from the analysis since their associated eigenvalues contain nothing but pure error. These three terms are related to the residual standard deviation, RSD, in the following way:

$$rc(\text{RSD})^2 = \sum_{i=1}^{r} \sum_{k=1}^{c} e_{ik}^2 \tag{4.35}$$

$$rn(\text{RSD})^2 = \sum_{i=1}^{r} \sum_{j=1}^{n} (\sigma_{ij}^{\dagger})^2 \tag{4.36}$$

$$r(c - n)(\text{RSD})^2 = \sum_{i=1}^{r} \sum_{j=n+1}^{c} (\sigma_{ij}^{0})^2 \tag{4.37}$$

Each of these expressions represents a different way of obtaining the residual standard deviation.

Placing (4.35), (4.36), and (4.37) into (4.34) and dividing through by rc, we conclude that

$$(\text{RSD})^2 = \frac{n}{c} (\text{RSD})^2 + \frac{c-n}{c} (\text{RSD})^2 \tag{4.38}$$

This important identity summarizes the theoretical arguments presented. The RSD can be interpreted[1,2] to be composed of two terms: *imbedded error* (IE) and *extracted error* (XE). In other words, the residual standard deviation, which is the *real error* (RE) can be expressed in a Pythagorean fashion as follows:

$$(RE)^2 = (IE)^2 + (XE)^2 \tag{4.39}$$

where

$$RE = RSD \tag{4.40}$$

$$IE = RSD \sqrt{\frac{n}{c}} \tag{4.41}$$

$$XE = RSD \sqrt{\frac{c - n}{c}} \tag{4.42}$$

The imbedded error is due to the fact that only a fraction of the error from the data mixes into the factor analytical reproduction process. This error becomes imbedded into the factors and cannot be removed by repeated factor analysis. The extracted error is due to the fact that some error is extracted when the secondary eigenvectors are dropped from the scheme.

The imbedded error (IE) is a measure of the difference between the pure data and the factor analysis reproduced data, whereas the real error (RE) is a measure of the difference between the pure data and the raw experimental data. As a mnemonic we can symbolically represent these statements in the form of a right triangle as shown in Figure 4.2.

Equation (4.41) shows that when n is less than c, IE is less than RE. Thus the error between the factor analysis-reproduced data and the pure data is less than the original error between the raw data and the pure data. Hence by using the primary abstract eigenvectors of factor analysis, we can always improve the data

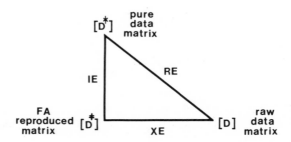

Fig. 4.2. Mnemonic diagram of the Pythagorean relationship between the theoretical errors [Reprinted with permission from E. R. Malinowski, *Anal. Chem.,* **49,** 606 (1977)].

Table 4.3 **Summary of results of factor-analyzing model data**[a]

		Artificial Data and Artificial Error			Factor Analysis of Raw Data		
n	$r \times c$	Range of Pure Data	Range of Error	True RMS of Errors	RE(RSD)	IE	RMS of $(d_{ik}^1 - d_{ik}^*)$
1	10×2	1 to 10	-0.2 to 0.2	0.148	0.170	0.120	0.087
2	16×5	2 to 170	-0.08 to 0.08	0.041	0.041	0.026	0.026
2	16×5	2 to 170	-0.99 to 0.91	0.566	0.511	0.323	0.381
2	15×5	0 to 27	-1.9 to 2.0	1.040	0.962	0.608	0.726
3	10×6	0 to 32	-0.9 to 0.9	0.412	0.376	0.266	0.311
4	16×9	-581 to 955	-1.0 to 1.0	0.548	0.499	0.333	0.398
5	10×9	-137 to 180	-1.0 to 1.0	0.463	0.372	0.277	0.400

[a] Reprinted with permission from E. R. Malinowski, *Anal. Chem.*, **49**, 606 (1977).

(even when we are not able to identify the true underlying factors) simply by having more than n columns in our data matrix.

At this point it is informative to point out that the root-mean-square (RMS) error, defined by (4.46) in a later section, is a measure of the difference between the factor analysis-reproduced data and the raw experimental data. A quite interesting fact now becomes apparent: the RMS is identical to the extracted error (XE). This can be verified by placing (4.44) into (4.42) and comparing the result with (4.52).

Malinowski[1,2] reported the results of an error study using artificially constructed data matrices. His findings are summarized in Table 4.3. The first column gives n, the number of factors used to generate a pure data matrix. The second column shows the number of rows and columns, $r \times c$, in each data matrix. The third column gives the range of error in the error matrix which was added to the pure data matrix to yield the raw data matrix. The "true RMS of the errors" (i.e., the difference between the raw data and the pure data) of the various error matrices are listed in the fifth column.

When these artificially generated raw data matrices were subjected to AFA, using the proper number of primary eigenvalues, the RE and IE values listed in the table were obtained. RE and IE were calculated from the abstract eigenvectors produced by the factor analysis by means of (4.44) and (4.41). Theoretically, the calculated RE should be identical to the true RMS of the errors. Comparing values in the fifth column with the corresponding values in the sixth column, we see that this is true (realizing that only the first digit of an error has significance).

The last column in the table contains the RMS of the difference between the FA reproduced data and the pure data. These values should be identical to the IE values calculated by the abstract eigenvalues. From the table we see that this is generally so within the first digit. Perfect agreement should not be expected because of the statistically small number of data points used in the model analyses.

Using an excessive number of eigenvectors in the reproduction process, thus minimizing the RMS error, is a practice that must be avoided because this tends to minimize the extracted error. Extra factors simply reproduce unwanted experimental error. Conversely, an insufficient number of eigenvectors will not account for the true variables at play. The RMS error will be too large. We must therefore search for various criteria to judge the exact number of factors. This problem is attacked in the next section.

4.3   DEDUCING THE NUMBER OF FACTORS

Deducing the exact size of the true factor space is a difficult problem because of experimental error. The various techniques developed to solve this problem can be divided into two general categories: (1) methods based upon a knowledge of the experimental error, and (2) approximate methods requiring no knowledge of the experimental error. Obviously, methods in the first category are preferred when the error is known. Often such information is lacking and methods in the second category must be employed because they offer a solution, albeit of a more dubious nature.

4.3.1   Methods Based on Experimental Error

Various criteria have been developed for determining the size of the true factor space when the experimental error is known. Many of these criteria are described in several excellent texts[3-6] written for the social scientist rather than the physical scientist. In order to compare one criterion with another we will, whenever possible, apply the criterion to the same data matrix chosen from the chemical literature. For this purpose we have selected the infrared absorbance matrix of Bulmer and Shurvell[7] to serve as an illustrative example. This matrix consists of 1800 data points concerning the absorbances of nine solutions of acetic acid in carbon tetrachloride measured at 200 different wavelengths in the carbonyl region. The standard deviation in absorbance was estimated to vary between 0.0005 and 0.0015 absorbance unit.

The rank of this absorbance matrix, according to Beer's law, will equal the number of absorbing components. Since acetic acid may form dimers, trimers, and so on, n is not necessarily equal to unity.

When the absorbance matrix was subjected to AFA, using covariance about the origin, nine characteristic eigenvectors resulted.[7] The corresponding eigenvalues are listed in Table 4.4 in the order of decreasing value. The question we will attempt to answer in this section is: Which of these eigenvectors belongs to the primary set and which belong to the secondary set? In other words, what is the true rank of the example data matrix?

Table 4.4 Illustrative example—factor analysis of the molar absorbances of the carbonyl region of solutions containing acetic acid in CCl_4[a]

n	Eigenvalue, (λ)	Standard Error in λ	Real Error, RE[b]	Imbedded Error, IE	Indicator Function, $IND \times 10^5$	Chi-Squared (χ^2)		
						Calculated	Expected[c]	3σ Misfit
1	19.193396	0.00438	0.015461	0.005154	24.16	363,995.8	1592	983
2	0.368079	0.00295	0.003208	0.001512	6.55	10,479.8	1386	298
3	0.009065	0.00470	0.002110	0.001406	5.86	5,504.9	1182	152
4	0.004414	0.00372	0.000962	0.000642	3.85	780.9	980	0
5	0.000294	0.00393	0.000889	0.000662	5.56	561.5	780	0
6	0.000260	0.00475	0.000787	0.000643	8.74	329.0	582	0
7	0.000141	0.00315	0.000760	0.000670	19.0	194.6	386	0
8	0.000132	0.00356	0.000704	0.000663	70.4	81.5	192	0
9	0.000099	0.00351	—	—	—	0.0	0	0

[a] Reprinted by permission from J. T. Bulmer and H. F. Shurvell, *J. Phys. Chem.,* **75,** 256 (1973) and from E. R. Malinowski, *Anal. Chem.,* **49,** 612 (1977).
[b] Also known as the residual standard deviation (RSD).
[c] χ^2(expected) $= (r - n)(c - n)$, where $r = 200$ and $c = 9$.

Residental Standard Deviation. As a guide to selecting the primary eigenvalues, many factor analysts use the residual standard deviation method. The residual standard deviation (RSD) is defined by the following equation:

$$r(c - n)\,(RSD)^2 = \sum_{i=1}^{r} \sum_{j=n+1}^{c} (\sigma_{ij}^0)^2 \tag{4.43}$$

The RSD in this expression is identical to the real error (RE) as defined by (4.40). Utilizing (4.17) we can rearrange (4.43) to read as follows:

$$RSD = \left[\frac{\sum_{j=n+1}^{c} \lambda_j^0}{r(c - n)} \right]^{1/2} \tag{4.44}$$

This equation is based upon "covariance about the origin." If we use "correlation about the origin" instead, we would obtain the following expression:

$$RSD = \left[\frac{\sum_{i=1}^{r} \sum_{k=1}^{c} d_{ik}^2 \sum_{j=n+1}^{c} \lambda_j^0}{rc(c - n)} \right]^{1/2} \tag{4.45}$$

These equations provide us with an excellent criterion for deducing which eigenvalues belong to the primary and secondary sets. To begin the deductive process we first calculate the RSD on the basis of only one factor. In this case the largest, hence most important, eigenvalue represents the primary eigenvalue. All the remaining eigenvalues belong to the secondary set and are included in the summation as indicated in (4.44). We then compare the RSD calculated from (4.44) to the estimated experimental error. If the RSD is greater than the estimated error, we have not chosen a sufficient number of factors. If the RSD is approximately equal to the estimated error, we have chosen the proper number of factors. The factor space would be one-dimensional.

On the other hand, if the RSD is greater than the estimated error, we then investigate the situation in which the two largest eigenvalues constitute the primary set. Once again we calculate the RSD by means of (4.44) and again compare it to the estimated error. If the RSD exceeds the estimated error, we repeat the process assuming that the three largest eigenvalues belong to the primary set. We continue in this manner, each time transferring the next largest eigenvalue from the secondary set to the primary set until we find that the RSD approximately equals the estimated error. When this occurs, we will have found which eigenvalues (and their associated eigenvectors) belong to the true primary and secondary sets.

As an example of this procedure, let us study the results of the illustrative example presented in Table 4.4. The RSD (RE) values were calculated from the eigenvectors by means of (4.44). Recalling that the error in the absorbances was estimated to be between 0.0005 and 0.0015, we conclude that there must be four recognizable factors because the largest RSD that does not exceed 0.0015 is 0.000962, corresponding to $n = 4$. For $n = 3$ the RSD (0.002110) is outside the acceptable range.

Because the residual standard deviation is an extremely useful concept in factor analysis, it is important for us to understand its full significance. If the errors are perfectly random, the error points will have a spherically symmetric distribution in the error space. Because of this distribution, the average of the sum the squares of the projections of the data points onto any secondary eigenvector axis should be approximately the same. Hence the RSD should be the same, regardless of how many error axes are employed in the calculation. In reality, the error distribution will not be perfectly symmetric. In fact, the principal component feature of FA actually searches for axes that exaggerate the effects of this asymmetry. RSD values calculated from PFA will, however, be approximately equal to the estimated error when the proper number of factors is employed. Principal factor analysis was used in obtaining the RSD values listed in Table 4.4.

Root-Mean-Square Error. The root-mean-square (RMS) error is defined by the equation

$$rc(\text{RMS})^2 = \sum_{i=1}^{r} \sum_{k=1}^{c} (d_{ik} - d_{ik}^\dagger)^2 \tag{4.46}$$

The term in parentheses on the right-hand side represents the difference between the raw data and the data regenerated by FA using only the primary set of eigenvectors. The sum is taken over all the data points. As strange as it may appear at first glance, this sum, involving the square of the difference, exactly equals the sum of the differences of the squares:

$$\sum_{i=1}^{r} \sum_{k=1}^{c} (d_{ik} - d_{ik}^{\dagger})^2 = \sum_{i=1}^{r} \sum_{k=1}^{c} (d_{ik}^2 - d_{ik}^{\dagger 2}) \tag{4.47}$$

Since this equality is by no means obvious, proof will be given here. First, we recall that

$$e_{ik}^0 = d_{ik} - d_{ik}^{\dagger} \tag{4.48}$$

We now make the following sequence of observations:

$$d_{ik}^2 = (d_{ik}^{\dagger} + e_{ik}^0)^2$$
$$= d_{ik}^{\dagger 2} + 2d_{ik}^{\dagger}e_{ik}^0 + e_{ik}^{02}$$
$$\sum\sum d_{ik}^2 - \sum\sum d_{ik}^{\dagger 2} = 2 \sum\sum d_{ik}^{\dagger}e_{ik}^0 + \sum\sum e_{ik}^{02}$$

The double sums are taken over $i = 1$ to $i = r$ rows and $k = 1$ to $k = c$ columns. But from (4.30) we learn that

$$\sum\sum d_{ik}^{\dagger}e_{ik}^0 = 0 \tag{4.49}$$

Therefore,

$$\sum\sum d_{ik}^2 - \sum\sum d_{ik}^{\dagger 2} = \sum\sum e_{ik}^{02} \tag{4.50}$$

Upon recalling (4.44), we find that

$$\sum\sum d_{ik}^2 - \sum\sum d_{ik}^{\dagger 2} = \sum\sum (d_{ik} - d_{ik}^{\dagger})^2 \tag{4.51}$$

which is what we wanted to prove.

By substituting (4.51) into (4.46) and recalling (4.27), we conclude that

$$\text{RMS} = \left[\frac{\sum_{j=n+1}^{c} \lambda_j^0}{rc} \right]^{1/2} \tag{4.52}$$

By comparing this result with (4.44), an analogous equation concerning the residual standard deviation, we see that

$$\text{RMS} = \left(\frac{c - n}{c} \right)^{1/2} (\text{RSD}) \tag{4.53}$$

Although RMS and RSD are closely related, they measure two entirely different errors. The root-mean-square-error calculation measures the difference between raw data and factor analysis regenerated data. The residual standard deviation measures the difference between raw data and pure data possessing no experimental error. For this reason caution must be exercised when applying and interpreting these errors. Since c is greater than n, from (4.53) we see that RMS is less than RSD. The use of the RMS error as a criterion for rank analysis can be misleading and is not recommended.

Average Error. Several investigators have used the *average error, $\bar{e}$,* as a criterion for fit. The average error is simply the average of the absolute values of the differences between the original and regenerated data. This average is compared to an estimated average error. Eigenvectors are systematically added to the scheme until the calculated average error approximately equals the estimated error.

This method, in reality, is analogous to the root-mean-square-error criterion, because, for a statistically large number of data points, the average error is directly proportional to the root-mean-square error:

$$\bar{e} = \left(\frac{2}{\pi}\right)^{1/2}(\text{RMS}) \tag{4.54}$$

Chi-Squared. Bartlett[8] proposed using a *chi-squared* (χ^2) criterion when the standard deviation varies from one data point to another and is not constant throughout the data matrix. This method takes into account the variability of the error from one data point to the next. Its disadvantage is that one must have a reasonably accurate error estimate for each data point. In this case χ_n^2 is defined as follows:

$$\chi_n^2 = \sum_{i=1}^{r} \sum_{k=1}^{c} \frac{(d_{ik} - d_{ik}^\dagger)^2}{\sigma_{ik}^2} \tag{4.55}$$

where σ_{ik} is the standard deviation associated with the measurable d_{ik}, $d_{ik}^\dagger$ is the value of the corresponding point regenerated from FA using the n largest eigenvalues, and the sum is taken over all experimental points. For each set of eigenvectors, χ_n^2 is compared to its expectation value given by the product

$$\chi_n^2(\text{expected}) = (r - n)(c - n) \tag{4.56}$$

The procedure for using the chi-squared criterion is as follows. First χ_1^2 is calculated from the data regenerated with the eigenvector associated with the largest eigenvalue. This value of chi-squared is then compared to the expectation value, $(r - 1)(c - 1)$. If χ_1^2 is greater than the expectation, the factor space is one-dimensional or greater. Next, χ_2^2 is calculated using the two largest eigenvalues. If χ_2^2 is greater than the expectation value $(r - 2)(c - 2)$, the factor space is two-dimensional or greater. This procedure, using the largest eigenvalues, is continued until χ_n^2 is less than its corresponding expectation value $(r - n)$

$(c - n)$. At this crossover point, the true n is estimated to be that which yields a χ_n^2 closest to its expectation value.

This behavior is illustrated by the bench mark (Table 4.4). In this case the standard deviations in absorbance were estimated for each data point, varying from 0.0005 to 0.0015. To calculate χ^2 this dispersion in the standard deviation was employed in (4.55). The crossover in χ_n^2 occurs between $n = 3$ (5504.9 > 1182) and $n = 4$ (780.9 < 980). Since 780.9 is closer to 980 than 5504.9 is to 1182, there are probably four factors. Another example of the chi-squared criterion is found in Section 7.1.3.

Standard Error in the Eigenvalue. Another method for deducing the factor space, based upon a statistical criterion for the "vanishing" of an eigenvalue, was developed by Hugus and El-Awady.[9] They showed that the standard error in an eigenvalue is related to the standard deviations of the data points. In particular, they have shown that

$$\sigma_m = \left[\sum_{j=1}^{c} \sum_{k=1}^{c} c_{mj}^2 c_{mk}^2 \sigma(Z)_{jk}^2 \right]^{1/2} \tag{4.57}$$

where σ_m is the standard error in the mth eigenvalue, c_{mj} and c_{mk} are the jth and kth components of the mth eigenvector [see (3.32)], and

$$\sigma(Z)_{jk}^2 = \begin{cases} \sum_{i=1}^{r} (d_{ij}^2 \sigma_{ik}^2 + d_{ik}^2 \sigma_{ij}^2) & \text{for } j \neq k \\ \sum_{i=1}^{r} 4 d_{ij}^2 \sigma_{ij}^2 & \text{for } j = k \end{cases}$$

where σ_{ij} is the error in d_{ij}.

This statistical criterion allows the inclusion of individual error, which may vary from one data point to the next. The dimensionality of the factor space is taken to be the number of eigenvalues that have values larger than their respective standard error. Applying this criterion to the bench-mark data (Table 4.4), we conclude that there are four primary eigenvectors, because the standard error in each eigenvalue is less than its eigenvalue for $n = 1$ to $n = 4$ but greater than its corresponding eigenvalue for $n = 5$ to $n = 9$. Another example of the use of this criterion is found in Section 7.1.3.

Distribution of Misfit. Another method for determining the dimensionality of the factor space involves studying the number of misfits between the observed and regenerated data as a function of the number of eigenvectors employed. A regenerated data point is classified as a misfit if its deviation from the observed value is three or more times greater than the standard deviation, σ, estimated from experimental information.

The bench-mark data (Table 4.4) contained 1800 data points. With three eigenvectors, 152 regenerated data points had misfits greater than 3σ. With four factors no misfits were greater than 3σ. Hence the factor space is four-dimensional. Another example of the 3σ misfit criterion is given in Table 7.3 in Section 7.1.3. A weak point of this method is the arbitrariness in deciding how many misfits can be tolerated.

Kankare[10] suggested that factor analysis be used to smooth the data by neglecting all data points with misfits greater than 3σ, and substituting for them their factor analysis regenerated values. The purpose of the smoothing is to remove excessive errors introduced by those points whose deviation can only be considered accidental. In this case care must be exercised not to choose too small a value for n. After smoothing, the adjusted data is factor-analyzed, hopefully yielding more reliable results. This process is somewhat risky because one might actually force the adjusted points to conform to a smaller factor space than that which truly exists, particularly so for points that are truly unique.

Residual Covariance Matrix. Ritter and coworkers[11] developed a method for choosing the number of factors, based upon examining the residual errors in the covariance matrix. First, the data matrix is subjected to the "covariance-about-the-mean" preprocessing procedure. Ritter and coworkers showed that the standard deviation, $e(z_{ij})$, in the covariance element z_{ij} can be calculated by means of the expression

$$e(z_{ij}) = \sigma(d_{ij}) \left[\sum_{k=1}^{r} (d_{kj}^2 + d_{ki}^2) \right]^{1/2} \qquad (4.58)$$

where $\sigma(d_{ij})$ is the square root of the variance of the preprocessed data point d_{ij}, the sum being taken over all r elements in the respective data columns. The number of factors is the smallest value of n for which all of the elements in the residual covariance matrix approximately equal their corresponding standard deviations, that is, when

$$[\mathcal{R}]_{\text{minimum } n} \simeq [e(z_{ij})] \qquad (4.59)$$

Reduced Error Matrix. A method for determining the rank of a matrix was devised by Wallace and Katz.[12] This method depends upon the construction of a companion matrix $[S]$, whose elements are the estimated errors of the data matrix $[D]$. The method involves reducing the data matrix by a series of elementary operations to an equivalent matrix whose diagonal elements are maximized and whose elements below the principal diagonal are all zero. The error matrix $[S]$ is continually transformed into an equivalent reduced error matrix during the reduction of $[D]$. This is done by a series of elementary operations based upon the theory of propagation of errors. The rank of the data matrix is equal to the number of diagonal elements in the reduced data matrix which are statistically nonzero. A diagonal element is arbitrarily considered to be nonzero if its absolute value is greater than three times the absolute value of the corresponding principal diagonal element of the reduced error matrix.

Interpretability of Target Tests. A promising method for finding the size of the factor space, without relying upon a knowledge of the error in the data matrix, involves interpreting the target test vectors. Employing too few eigenvectors may lead to a poor fit between a predicted vector and its corresponding test vector, whereas employing too many eigenvectors will tend to reproduce experimental error and the agreement will be overly good. Unfortunately, to apply this method we need to know how much error can be tolerated in the test vector.

Another method for determining n from target testing involves comparing the predicted values with known values for a free-floated test point as a function of the number of factors employed. The true factor space is identified when the predicted values agree, within experimental error, with the known values. At the present time, the role of error in the target test is too poorly understood to make this approach a viable criterion.

These target methods are included in this section because they do require a knowledge of the error in the target.

Conclusion. The methods described in this section for determining the real factor space of a data matrix depend upon an accurate estimate of the error. Each method may lead to a different conclusion. This problem was investigated by Dueweriet al.,[13] who concluded that the determination of the true rank of data having uncertainty is not a trivial task and that none of the various rank determination criteria is clearly superior or completely satisfactory when used alone. The various criteria, examined together, afford a better guide than reliance on a single rule.

Duewer and coworkers based their conclusions on a manufactureed set of data based on the dimensions of a rectangular solid, involving the following parameters: length (L), width (W), and thickness (T); three wraparound lengths $2(L + W)$, $2(L + T)$, and $2(W + T)$; and the three diagonals $(L^2 + W^2)^{1/2}$, $(L^2 + T^2)^{1/2}$, and $(W^2 + T^2)^{1/2}$. Random digits were generated for L, W, and T and a true (pure) data matrix, consisting of the nine variables above, was formed from 25 sets of L, W, and T. To test the effects of analytical uncertainty, errors were added to the data and studies were conducted on both the true data matrix and the perturbed data matrix. Various error criteria (such as χ^2 and $\bar{e}$) for judging the rank of the perturbed data matrix were compared to the results obtained from the true data matrix. Data pretreatment methods such as C_o, C_m, R_o, and R_m (see Section 3.2.3) were also investigated.

The effects of adding artificial error to chemical data matrices have also been investigated by others.[14,15] Because these studies are quite detailed, interested readers are referred to the original papers.

4.3.2 Methods Requiring No Knowledge of Experimental Error

In the previous section we studied various methods for determining the factor space when accurate estimates of the errors are known. Often these methods cannot be used because such information is either not available or is highly suspect. A more difficult problem is to deduce the factor space without relying upon an estimation of the error. In this section we explore various methods that have been proposed to solve this challenging problem. In Section 4.4 we learn how these methods can be used to deduce not only the size of the factor space but also the size of the error.

Imbedded Error Function. The *imbedded error function* can be used to determine the number of factors in a data matrix without relying upon any estimate of the error.[1,16] By inserting (4.44) into (4.41), we see that

$$\mathrm{IE} = \left[\frac{n \sum\limits_{j=n+1}^{c} \lambda_j^0}{rc(c-n)} \right]^{1/2} \tag{4.60}$$

The imbedded error is a function of the secondary eigenvalues, the number of rows and columns in the data matrix, and the number of factors. Because this information is always available to us when we perform factor analysis, we can calculate IE as a function of n, as n goes from 1 to c. By examining the behavior of the IE function as n varies, we can often deduce the true number of factors. The IE function should decrease as we use more and more primary eigenvectors in data reproduction. However, once we have exhausted the primary set and begin to include secondary eigenvectors in the reproduction, the IE should increase. This should occur because a secondary eigenvalue is simply the sum of the squares of the projections of the error points onto an error axis. If the errors are distributed uniformly, their projections onto each error axis should be approximately the same. In other words, $\lambda_j^0 \simeq \lambda_{j+1}^0 \simeq \lambda_c^0$ and (4.60) becomes

$$\mathrm{IE} \simeq n^{1/2}k \qquad \text{for} \quad n > \text{true } n \tag{4.61}$$

In (4.61), k is a constant:

$$k = \left[\frac{\lambda_j^0}{rc} \right]^{1/2} \tag{4.62}$$

These equations apply only when we have utilized an excessive number of eigenvectors in the reproduction process. Equation (4.61) shows that the IE will actually increase once we begin to use more factors than the true number required. In practice, a steady increase in IE will rarely be observed because the principal component feature of FA exaggerates the nonuniformity in the error distribution. Hence the secondary eigenvalues will not be exactly equal. The minimum in the IE function will not be clearly defined if the errors are not fairly uniform throughout, if the errors are not truly random, if systematic errors exist, or if sporadic errors exist.

Three examples illustrating the behavior of the IE function,[16] taken from chemical problems, are given next.

Example 1:   Results of factor-analyzing the infrared absorbances of acetic acid in CCl_4 solutions are given in Table 4.4. Here the IE shows a progressive decrease on going from $n = 1$ to $n = 4$, but shows no further decrease on going from $n = 4$ to $n = 8$. This gives evidence that four factors are responsible for the absorbance matrix. This conclusion is consistent with the other approaches described in Section 4.3.1.

Example 2:   Table 7.1 concerns the factor analysis of the spectrophotometric absorbances of 38 solutions of $[(en)_2Co(OH)_2Co(en)_2]^{4+}$ measured at nine different wavelengths. The fact that the IE reaches a minimum at $n = 3$ gives evidence that there are three species responsible for the absorbance.

Example 3:   Table 4.5 shows the results of factor-analyzing the gas–liquid-chromatographic retention indices of 22 ethers on 18 chromatographic columns. We see here that no minimum appears in the IE function. This misbehavior could be caused by any one or combination of the four reasons discussed

Table 4.5   Results of factor-analyzing the GLC retention indices of 22 ethers on 18 chromatographic columns[a]

n	RE	IE	IND	n	RE	IE	IND
1	22.28	5.25	0.07708	10	1.40	1.04	0.02187
2	7.25	2.42	0.02831	11	1.25	0.98	0.02553
3	5.30	2.16	0.02354	12	1.07	0.87	0.03975
4	4.06	1.91	0.02070	13	0.94	0.80	0.03748
5	3.24	1.71	0.01915	14	0.73	0.65	0.04586
6	2.76	1.59	0.01914	15	0.69	0.63	0.07618
7	2.42	1.51	0.01997	16	0.61	0.58	0.15261
8	2.05	1.36	0.02045	17	0.59	0.57	0.59012
9	1.71	1.21	0.02114	—	—	—	—

[a] Reprinted with permission from E. R. Malinowski, *Anal. Chem.*, **49**, 612 (1977).

earlier. Unfortunately, in this case we cannot use the IE function to determine the number of factors present.

Factor Indicator Function. Malinowski[1,16] discovered an empirical function, called the *factor indicator function,* which appears to be much more sensitive than the IE function in its ability to pick out the proper number of factors. The factor indicator function (IND) is defined as

$$\text{IND} = \frac{\text{RE}}{(c - n)^2} \tag{4.63}$$

This function is composed of exactly the same variables as IE, namely λ_j^0, r, c, and n. The IND function, similar to the IE function, reaches a minimum when the correct number of factors are employed. The minimum, however, is much more pronounced and, more important, often occurs in situations in which the IE exhibits no minimum.

To see the behavior of the IND function, let us examine the same three examples studied in the previous subsection. The first example concerns the absorbances of acetic acid in CCl_4, Table 4.4. In this table we see that the IND function reaches a minimum at $n = 4$, in agreement with our conclusions based on IE. In the second example, the IND function for the cobaltic solutions, shown in Table 7.1, reaches a minimum at $n = 3$. This is also in agreement with our findings based upon the IE function. The third example concerns the GLC retention indices. In Table 4.5, a minimum in the IND is clearly present at $n = 6$, whereas no minimum appears in the IE function. Apparently, the IND function somehow compensates for the principal component exaggeration of nonuniformity of the error distribution. The IND function is not fully understood at the present time, and it should be used with caution.

Variance. The factor analytical solution can be expressed as a linear sum of outer products of row-designee vectors and column-designee vectors:

$$[D] = R_1 C_1' + R_2 C_2' + \cdots + R_j C_j' + \cdots + R_c C_c' \tag{4.64}$$

Here R_j and C_j' are the row-designee and column-designee vectors associated with eigenvalue λ_j. Each outer product yields an $r \times c$ matrix, $[D_j]$, which accounts for a portion of the raw data matrix, so that

$$[D] = [D_1] + [D_2] + \cdots + [D_j] + \cdots + [D_c] \tag{4.65}$$

where

$$[D_j] = R_j C_j' \tag{4.66}$$

In other words, each $[D_j]$, which is associated with a particular eigenvector, C_j, and a particular eigenvalue, $\lambda_j = R_j' R_j$ [see (3.68)], accounts for some of the variance in the raw data matrix. The *variance* is defined as follows:

$$\text{variance} = \frac{\displaystyle\sum_{i=1}^{r}\sum_{k=1}^{c} d_{ik}^2(j)}{\displaystyle\sum_{i=1}^{r}\sum_{k=1}^{c} d_{ik}^2} \tag{4.67}$$

where $d_{ik}(j)$ is an element of $[D_j]$ and d_{ik} is an element of $[D]$, the raw data matrix. The sums are taken over all $r \times c$ elements of the matrices. Based upon arguments given in Section 4.1.1, one can easily show that the variance can be calculated directly from the eigenvalues; that is,

$$\text{variance} = \frac{\lambda_j}{\displaystyle\sum_{j=1}^{c} \lambda_j} \tag{4.68}$$

Because the variance measures the importance of an eigenvector, it can be used as a criterion for accepting or rejecting an eigenvector. In practice, eigenvectors having large variances are considered to be primary eigenvectors, whereas eigenvectors having small variances are considered to be secondary eigenvectors. Unfortunately, classifying the variance as large or small presents a problem. It is at this critical point in the process that various investigators diverge. Often, the factor analyst gives no real justification for his decision of the cutoff point used in the variance classification, thus casting doubt on the final decision.

Cumulative Percent Variance. The cumulative percent variance is a measure of the percent of the total variance in the data which is accounted for by short-circuit reproduction. It is defined as follows:

$$\text{cumulative percent variance} = 100\left(\frac{\sum\sum d_{ik}^{\dagger 2}}{\sum\sum d_{ik}^2}\right) \tag{4.69}$$

Here $d_{ik}^{\dagger}$ is the value of a data point reproduced by AFA and d_{ik} is the raw, experimental data point. The sums are taken over all data points. Inserting (4.24) and (4.26) into (4.69), we see that the cumulative percent variance can be expressed in terms of the eigenvalues:

$$\text{cumulative percent variance} = 100\left(\frac{\displaystyle\sum_{j=1}^{n} \lambda_j^{\dagger}}{\displaystyle\sum_{j=1}^{c} \lambda_j}\right) \tag{4.70}$$

For this reason the method is often referred to as the *percent variance in the eigenvalue.*

The percent variance criterion accepts the set of largest eigenvalues required to account for the variance within a chosen specification. The problem then becomes one of estimating exactly how much variance in the data must be accounted for. Arbitrary specifications such as 90%, 95%, or 98% variance do not provide reliable estimates for judging the number of factors. In general, the method can be deceptively misleading and is not recommended unless one can make an accurate estimate of the true variance in the data. In practice, this cannot be done without a knowledge of the error.

Scree Test. The *Scree test,* proposed by Cattell,[17] is based on the observation that the residual variance should level off before the factors begin to account for random error. The residual variance is defined as follows:

$$\text{residual variance} = \frac{\sum\limits_{i=1}^{r} \sum\limits_{k=1}^{c} (d_{ik}^2 - d_{ik}^{\dagger 2})}{rc} \tag{4.71}$$

From (4.46) and (4.47) we see that the residual variance is equal to the square of the RMS error, which can be calculated from the secondary eigenvalues by means of (4.52). The residual percent variance is defined as

$$\text{residual percent variance} = 100 \left[\frac{\sum\limits_{i=1}^{r} \sum\limits_{k=1}^{c} (d_{ik}^2 - d_{ik}^{\dagger 2})}{\sum\limits_{i=1}^{r} \sum\limits_{k=1}^{c} d_{ik}^2} \right] \tag{4.72}$$

In terms of the eigenvalues, this expression takes the form

$$\text{residual percent variance} = 100 \left(\frac{\sum\limits_{j=n+1}^{c} \lambda_j^0}{\sum\limits_{j=1}^{c} \lambda_j} \right) \tag{4.73}$$

When the residual percent variance is plotted against the number of factors used in the reproduction, the curve should drop rapidly and level off at some point. The point where the curve begins to level off, or where a discontinuity appears, is used to deduce the factor space. This is illustrated in Figure 4.3, taken from the work of Rozett and Petersen,[18] who investigated the mass spectra data of the 22 isomers of $C_{10}H_{14}$. These investigators used covariance about the mean, C_m; covariance about the origin, C_o; correlation about the mean, R_m; and correlation about the origin, R_o (see Section 3.2.3). They found that leveling is sensitive to the technique employed. For R_m and C_m, information concerning the zero point of the experimental scale is lost. For R_o and R_m, information concerning the relative sizes of the data points is lost. Such information is retained only in the C_o technique, which is the procedure recommended by Rozett and Petersen. For C_o, the leveling occurs at $n = 3$.

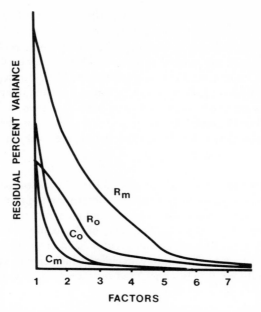

Fig. 4.3. Residual percent variance as a function of the number of factors employed in data reproduction. See Section 3.2.3 for definitions of C_m, C_o, R_o, and R_m [Reprinted with permission from R. W. Rozett and E. Petersen, *Anal. Chem.*, **47**, 1303 (1975)].

Average Eigenvalue. The *average eigenvalue criterion* proposed by Kaiser[19] has gained wide popularity among factor analysts. This criterion is based upon accepting all eigenvalues with values above the average eigenvalue and rejecting all those with values below the average. If one uses correlation about the origin, so that the correlation matrix has unities for each diagonal element, the average eigenvalue will be unity. In this case the criterion accepts all eigenvalues above unity, the average value, and excludes all eigenvalues below unity. This procedure is more popularly known as the *eigenvalue-one criterion*.

Exner Function. Weiner and coworkers[20] suggested that the *Exner* psi (ψ) *function* be employed when a good measure of the experimental error is not available. This function is defined by

$$\psi = \left[\frac{\sum\limits_{i=1}^{r} \sum\limits_{k=1}^{c} (d_{ik} - d_{ik}^{\dagger})^2}{\sum\limits_{i=1}^{r} \sum\limits_{k=1}^{c} (d_{ik} - \bar{d})^2} \times \frac{rc}{(rc) - n} \right]^{1/2} \tag{4.74}$$

Here d represents the grand mean of the experimental data. The ψ function can vary from zero to infinity, with the best fit approaching zero. A ψ equal to 1.0 is the upper limit of physical significance, because this means that one has not done any better than simply guess that each point has the same value of the grand mean. Exner[21] proposed that 0.5 be considered the largest acceptable ψ value, because this means that the fit is twice as good as guessing the grand mean for each point. Following Exner's reasoning, $\psi = 0.3$ is considered a fair correlation, $\psi = 0.2$ is considered a good correlation, and $\psi = 0.1$ is considered an excellent correlation. This method can only be expected to give a very crude estimate of the factor space.

Loading Distribution. Cattell[22] suggested that the factor loadings (i.e., the components of each eigenvector) be used to determine the factor space. Since each successive eigenvector involves more random error, the loadings should tend toward a normal distribution. The mean of the loadings of an eigenvector should be close to zero only if the eigenvector is an error eigenvector (i.e., a secondary eigenvector). Unfortunately, deciding which eigenvectors have mean loadings sufficiently small to be considered zero is arbitrary. Consequently, the method is subject to personal bias.

4.4   DEDUCING THE EXPERIMENTAL ERROR

In Section 4.3.2 we learned how various criteria can be used to determine the number of factors. Once we know the true number of factors, we can calculate the real error (RE) in the data matrix without relying upon any prior knowledge of the experimental error. This can be done because (4.44) and (4.40) lead to the expression

$$
RE = \left[\frac{\sum_{j=n+1}^{c} \lambda_j^0}{r(c-n)} \right]^{1/2}
\tag{4.75}
$$

where the required quantities λ_j^0, r, c, and n are all known. These same quantities are involved in the IE and IND functions, which can be used to determine the number of factors [see (4.60) and (4.63)].

To see how these error functions can be used jointly, let us return to the three examples cited in Section 4.3.2. For the infrared absorbances of acetic acid (example 1), the IE and IND functions leads us to surmize that $n = 4$ (see Table 4.4). From this we conclude that the RE is 0.000962, as shown in the table. This is in agreement with the estimate that the experimental error lies between 0.0005 and 0.0015 absorbance unit. For the absorbances of the cobalt complexes (example 2), we found $n = 3$. Applying (4.75), we find that RE $= 0.00104$, in agreement with the known facts. For the GLC analysis (example 3) we concluded that six factors were present (see Table 4.5). The RE corresponding to $n = 6$ is 2.76, in excellent agreement with the fact that the experimental error was estimated to be no greater than 3 units. Many other examples of the usage of the IE, IND, and RE functions can be found in the paper of Malinowski.[16]

4.5 TESTING FACTOR ANALYZABILITY

To be truly factor-analyzable, a data matrix must obey the linear-sum-of-products rule described by (1.1), must consist of a sufficient number of data columns, and must be free from excessive error. Malinowski[16] showed how the IE and IND functions can be used to decide whether a data matrix is factor-analyzable. He studied a series of data matrices composed of random numbers. An example of the type of results obtained is shown in Table 4.6. The IE increases on going from $n = 1$ to $n = 2$ to $n = 3$, indicating, theoretically, that we are dealing with pure error space. Also, the IND increases from $n = 1$ to $n = 7$. Since this matrix is obviously not one-dimensional, the minimum must occur at $n = 0$. If both the IE and IND functions behave in this fashion, the data are probably not factor-analyzable. By studying the behavior of the IE and IND functions, we can weed out those data matrices that are not factor-analyzable. This weeding-out process is an important adjunct to factor analytical studies and should be performed routinely. There exists, at the present time, no other method for accomplishing this important task.

Table 4.6 Results of factor-analyzing a 10 × 8 data matrix composed of random numbers ranging from 4 to 99 [a]

n	RE	IE	IND
1	27.00	9.54	0.55
2	24.66	12.33	0.69
3	21.88	13.40	0.88
4	18.51	13.09	1.16
5	14.42	11.40	1.60
6	11.02	9.54	2.75
7	6.47	6.05	6.47

[a] Reprinted with permission from E. R. Malinowski, *Anal. Chem.,* **49**, 612 (1977).

4.6   ERRORS IN TARGET TEST VECTORS

In this section we will trace through the target testing process to develop error criteria that can be used to determine whether a test vector is a valid factor and whether the test vector will lead to data matrix improvement or degradation.

4.6.1   Theory

Arguments presented in Section 4.1 form the basis of the approach.[23] There we learned how the experimental errors in the raw data matrix perturb the abstract cofactors. In fact, (4.9) and (4.21) show exactly how the errors contribute to the row and column cofactors, respectively. If the errors in the data matrix are reasonably small, the ratio c^*_{jk}/c_{jk} will be close to unity and (4.9) will reduce to the following:

$$r_{ij} = r^*_{ij} + \sigma^{\ddagger}_{ij} \qquad (4.76)$$

From this we conclude that the abstract row-cofactor matrix, $[R^{\ddagger}]$, can be expressed as a sum of two matrices: $[R^*]$, the pure row matrix, and $[E^{\ddagger}]$, the associated error matrix:

$$[R^{\ddagger}] = [R^*] + [E^{\ddagger}] \qquad (4.77)$$

We recall, from (3.97), that the elements of $[R^{\ddagger}]$ are used in conjunction with the eigenvalues to yield a least-squares transformation vector, T_l, for a given test vector, $\overline{\overline{R}}_l$. The predicted vector, $\overline{R}_l$, is calculated from these results by means of (3.84).

$$\overline{R}_l = [R^{\ddagger}]T_l \qquad (3.84), (4.78)$$

The *apparent error in the target test vector, E_A,* is defined as the difference between the "predicted" vector, $\overline{R}_l$ and the "raw" test vector, $\overline{\overline{R}}_l$,

$$E_A = \overline{R}_l - \overline{\overline{R}}_l \qquad (4.79)$$

Placing (4.77) into (4.78) and then inserting the result into (4.79), we find that

$$E_A = [E^{\ddagger}]T_l + [R^*]T_l - \overline{\overline{R}}_l \qquad (4.80)$$

If the test vector represents a true factor, and if both the test vector and the data matrix are pure (i.e., contain no experimental error), then

$$\overline{\overline{R}}^*_l = \overline{R}^*_l = [R^*]T^*_l \qquad (4.81)$$

Here $\overline{\overline{R}}^*_l$ is the pure test vector, $\overline{R}^*_l$ the pure predicted vector, and T^*_l the transformation vector obtained using pure data. Such situations almost never exist in chemistry. If, however, the errors are small, we might expect

$$\overline{\overline{R}}_l^* \simeq [R*]T_l \tag{4.82}$$

where T_l is the transformation vector obtained using raw data. Hence

$$E_A = [E^\ddagger]T_l + \overline{\overline{R}}_l^* - \overline{\overline{R}}_l \tag{4.83}$$

The *real error in the target test vector*, E_T, is defined as the difference between the pure test vector and the raw test vector:

$$E_T = \overline{\overline{R}}_l^* - \overline{\overline{R}}_l \tag{4.84}$$

The *real error in the predicted vector*, E_p, is defined as the difference between the predicted vector and the pure test vector:

$$E_p = \overline{R}_l - \overline{\overline{R}}_l^* \tag{4.85}$$

From the equations above we see that

$$E_A = E_p + E_T \tag{4.86}$$

and

$$E_p = [E^\ddagger]T_l \tag{4.87}$$

It is better to express the errors as root mean squares rather than as vectors. This is easily accomplished because the inner product of an error vector is related to its standard deviation. For example, the root mean square of the apparent error in the test vector (AET) is defined by

$$\text{AET} = \left[\frac{\sum\limits_{i=1}^{r} (\bar{r}_i - \overline{\overline{r}}_i)^2}{r} \right]^{1/2} \tag{4.88}$$

Here $\bar{r}_i$ and $\overline{\overline{r}}_i$ are the ith elements of the predicted vector and the test vector, respectively. The sum, taken over all r elements of the vector, is equal to the dot product of error vector E_A:

$$E_A^T \cdot E_A = \sum\limits_{i=1}^{r} (\bar{r}_i - \overline{\overline{r}}_i)^2 \tag{4.89}$$

Hence from (4.88) and (4.89) we see that

$$E_A^T \cdot E_A = r(\text{AET})^2 \tag{4.90}$$

The root mean square of the predicted vector (REP) is defined by

$$\text{REP} = \left[\frac{\sum\limits_{i=1}^{r} (\bar{r}_i - \overline{\overline{r}}_i^*)^2}{r} \right]^{1/2} \tag{4.91}$$

Here $\bar{\bar{r}}_i^*$ is the ith element of the pure test vector. Realizing that

$$E_p^T \cdot E_p = \sum_{i=1}^{r} (\bar{r}_i - \bar{\bar{r}}_i^*)^2 \tag{4.92}$$

and using (4.91) and (4.92), we conclude that

$$E_p^T \cdot E_p = r(\text{REP})^2 \tag{4.93}$$

Similarly, in standard RMS form, the real error in the target vector (RET) is defined as

$$\text{RET} = \left[\frac{\sum_{i=1}^{r} (\bar{\bar{r}}_i^* - \bar{\bar{r}}_i)^2}{r} \right]^{1/2} \tag{4.94}$$

Noting that

$$E_T^T \cdot E_T = \sum_{i=1}^{r} (\bar{\bar{r}}_i^* - \bar{\bar{r}}_i)^2 \tag{4.95}$$

and using (4.94) and (4.95), we find that

$$E_T^T \cdot E_T = r(\text{RET})^2 \tag{4.96}$$

With this information we can convert (4.86), which shows the relationship between the error vectors, into an RMS relationship. To do so, we consider the inner product of both sides of (4.86):

$$E_A^T \cdot E_A = E_p^T \cdot E_p + E_T^T \cdot E_T + E_p^T \cdot E_T + E_T^T \cdot E_p \tag{4.97}$$

Since the elements of these vectors are random errors, being both positive and negative, the last two terms on the right should be relatively small in comparison with the first two terms, which are sums of squares. Hence as an excellent approximation, we may write

$$E_A^T \cdot E_A = E_p^T \cdot E_p + E_T^T \cdot E_T \tag{4.98}$$

Substituting (4.90), (4.93), and (4.96) into (4.98), we conclude that

$$(AET)^2 = (REP)^2 + (RET)^2 \tag{4.99}$$

Figure 4.4 is a mnemonic representation of this Pythagorean relationship. The three different vectors ($\bar{R}$, $\bar{R}^*$, and $\bar{\bar{R}}$) concerning a single factor lie at the corners of the right triangle with $\bar{\bar{R}}^*$ coincident with the right angle. RET and REP are the sides of the triangle and AET is the hypothenuse.

Often, the apparent error in the target test vector cannot be obtained from (4.88) directly because the test vector may be incomplete, having only p elements instead of r. Applying statistical arguments based upon the number of degrees of freedom, we may write

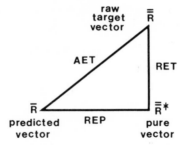

Fig. 4.4. Mnemonic diagram of the Pythagorean relationship of the errors in the target test vector [Reprinted with permission from E. R. Malinowski, *Anal. Chim. Acta,* **103,** 339 (1978)].

$$\left[\frac{\sum\limits_{i=1}^{r} (\bar{r}_i - \bar{\bar{r}}_i)^2}{r - n} \right] \simeq \left[\frac{\sum\limits_{i=1}^{p} (\bar{r}_i - \bar{\bar{r}}_i)^2}{p - n} \right] \tag{4.100}$$

Recall that n is the true number of factors. Inserting (4.100) into (4.88) gives us a more general expression for the apparent error:

$$\text{AET} = \left[\frac{r - n}{p - n} \times \frac{\sum\limits_{i=1}^{p} (\bar{r}_i - \bar{\bar{r}}_i)^2}{r} \right]^{1/2} \tag{4.101}$$

This expression is applicable for situations where the target is either complete or incomplete.

To obtain a numerical value for REP, we make the following observations. The inner product of E_p, according to (4.87), can be written in terms of the error matrix $[E^{\ddagger}]$ and the transformation vector T_l:

$$E_p^T \cdot E_p = \{[E^{\ddagger}] T_l\}^T \{[E^{\ddagger}] T_l\} = T_l^T [E^{\ddagger}]^T [E^{\ddagger}] T_l \tag{4.102}$$

Since $[E^{\ddagger}]$ is an $r \times n$ matrix, premultiplication by its own transpose will yield an $n \times n$ matrix with a trace equal to $rn(\text{RE})^2$, where RE is the real error in the data matrix [see (4.36) and (4.40)]. In comparison to the diagonal elements, the off-diagonal elements of this $n \times n$ matrix should be negligibly small because the elements of $[E^{\ddagger}]$ are error points randomly scattered about zero. Because there are n elements along the diagonal, the average value of a diagonal element is $r(\text{RE})^2$. Hence as a reasonable approximation, we can write

$$E_p^T \cdot E_p = r(\text{RE})^2 (T_l \cdot T_l) \tag{4.103}$$

where $(T_l \cdot T_l)$ is the dot product of the transformation vector. Inserting this into (4.93), we find that

$$REP = (RE) \sqrt{T_l \cdot T_l} \qquad (4.104)$$

Equation (4.104) affords an easy method for calculating the real error in the predicted vector. A numerical value for RE is obtained from the eigenvalues during the abstract reproduction step, which also yields the size of the factor space. When a test vector is target-transformed, we obtain numerical values for the transformation vector. These values are then placed into (4.104) and a numerical value for REP is obtained.

From (4.101) and (4.104) we can calculate numerical values for AET and REP, respectively. Inserting these into (4.99), we can compute RET, the real error in the test vector. This error can be compared to our estimation of the error in the data used in constructing the test vector. If the calculated RET reasonably agrees with the estimated RET, we would conclude that the target vector is valid. In the next two sections we develop empirical criteria for judging the validity of a test vector. Because of the approximations made, the derived error expressions will be valid only if the test vector is a true vector having sufficient accuracy. If the test vector is not a true representation of a valid factor, the resulting transformation vector will be incorrect, (4.82) will fail, and, consequently, all equations thereafter will be invalid. In the next section we develop empirical criteria to determine when this will occur.

4.6.2   Reliability Function

To judge the validity of a target vector, we can make use of the *reliability function* (RELI), defined as follows:[23]

$$RELI = \left(1 - \frac{(RET)^2 - (RET)_{est}^2}{(AET)^2}\right)^{1/2} \qquad (4.105)$$

Here $(RET)_{est}$ is the value of the real error in the target estimated directly from a knowledge of the experimental error in the data used to construct the test vector. An advantage of the RELI function lies in the fact that it involves a relative comparison rather than an absolute comparison.

Ideally, if the calculated RET is found to equal the estimated RET, the RELI would be unity, indicating 100% reliability. However, because of the approximations employed in deriving RET, we will often obtain RELI values which are greater than unity. In such cases we can be reasonably confident that the test vector is a true factor.

Conversely, if the calculated RET is much larger than the estimated RET, the RELI will be small and we would have little or no confidence that the associated test vector is a true factor. As a rule of thumb, the target is considered reliable if its RELI is 0.5 or greater, and unreliable if its RELI is less than 0.5. The RELI function gives us a sense of how close the target emulates a true factor.

4.6.3   Spoil Function

In many situations, owing to lack of information concerning the experimental uncertainty, the error in the test vector cannot be accurately estimated. Even when error information is available, a direct comparison between the calculated and estimated RET does not give a clear indication of the errors that will be introduced into the reproduced data matrix by target combination. In addition to determining whether or not a test vector is acceptable as a true factor, we need some criterion to tell us whether a given test vector will improve or spoil the reproduced data matrix when that vector is used in the TFA combination step. The theory in Section 4.6.1 affords us with the necessary tools to develop such a criterion, called the *SPOIL*.[23]

Before defining the SPOIL function, let us examine (4.99) and Figure 4.4. If a given target vector were pure, having no error, RET would be zero and AET would equal REP. The predicted vector, obtained by the method of least squares, would be different from the pure test vector. In this situation the error in the predicted vector comes solely from the error in the data matrix. Thus

$$REP \simeq EDM \tag{4.106}$$

where EDM is the error contributed by the data matrix. This approximation holds even when RET $\neq$ 0. Hence the apparent error in the target can be interpreted as a vector sum of two error sources: an error from the impure data matrix and an error from the impure target. Although somewhat oversimplified, this interpretation is conceptually satisfying.

If the real error in the target is less than the error contributed by the data matrix (i.e., if RET < EDM $\simeq$ REP), the reproduced target will contain more error than the original target. When a good target is used in the combination step it will tend to improve the regenerated data matrix. There is, therefore, a need to seek the most accurate data to describe the target.

On the other hand, if the real error in the target is greater than the error contributed by the data matrix (i.e., if RET > EDM $\simeq$ REP), the reproduced target vector will be more accurate than the original impure target vector. Thus target testing can be used as a unique data improvement tool. This is true only if the target is a real factor and if the correct number of factors are employed in the transformation test. Unfortunately, however, such a target, when used in the combination step, tends to spoil the reproduced data matrix because it introduces additional error into the reproduction process.

The SPOIL function[23] is defined as

$$\text{SPOIL} = \frac{\text{RET}}{\text{EDM}} \simeq \frac{\text{RET}}{\text{REP}} \qquad (4.107)$$

If the SPOIL is less than 1.00, the combination-reproduced data matrix will be improved by the target vector. On the other hand, if the SPOIL is greater than 1.00, the reproduced data matrix will be spoiled by the target. In fact, the larger the SPOIL, the poorer will be the data matrix reproduction. Consequently, the errors in the target test points should be sufficiently small so that negligible error is introduced into the FA process. This means that we should strive for targets that yield the smallest possible SPOIL.

The value of the SPOIL also provides the factor analyst with an excellent criterion for judging the overall validity of a suspected target. Since EDM is approximately equal to REP and since the equations used to evaluate RET and REP are not exact, we can only hope to develop rule-of-thumb criteria. By studying model sets of data, Malinowski[23] classified a target, on the basis of its SPOIL, as being acceptable, moderately acceptable, or not acceptable. A target is considered to be acceptable if its SPOIL lies between 0.0 and 3.0. If the SPOIL lies between 3.0 and 6.0, the target is moderately acceptable; although such a target is probably a true factor, the error is too large and its use in data matrix reproduction will lead to spoilage. A target is not acceptable if its SPOIL is greater than 6.0, because it will introduce excessive error into the reproduced data matrix.

4.6.4   Chemical Example

To illustrate the principles and the usefulness of target error theory in chemistry, Malinowski[23] investigated the solvent dependence of F^{19} magnetic resonance chemical shifts. When the data matrix, which consisted of the fluorine shifts of 19 rigid, nonpolar solute molecules dissolved in eight solvents, was subjected to AFA, three factors were found. This conclusion was based upon the IE and IND functions, which yielded an RE equal to 0.035 ppm, in excellent agreement with the known uncertainty.

Based upon theoretical arguments, the gas-phase shift of the solute was suspected to be one of the three basic factors. To see whether or not this was so, a gas-phase test vector, shown in Table 4.7, was constructed. This test vector contained only 12 points; seven points had to be free-floated because their gas-phase shifts were not measured. The predicted vector yielded gas-phase shifts for all 19 solutes, including the seven free-floated solutes. The AET, shown at the bottom of the table, was calculated from the differences between the test and predicted points, via (4.101). The table also lists values for REP and RET, calculated by means of (4.104) and (4.99), respectively.

Table 4.7   Target-testing ^{19}F gas-phase shifts[a]

Solute	Test Vector, $\underset{=}{\overline{R}}$	Predicted Vector, $\overline{R}$	Difference, $\overline{R} - \underset{=}{\overline{R}}$
CF_2Br_2	−2.27	−2.50	−0.23
$CFCl_3$	—	4.98	—
CF_2ClBr	4.64	4.52	−0.12
$CFCl_2CFCl_2$	—	71.62	—
sym-$C_6F_3Cl_3$	—	118.98	—
CF_2Cl_2	12.17	11.95	−0.22
cis-CFClCFCl	111.91	112.14	0.23
trans-CFClCFCl	—	125.62	—
C_6F_6	170.73	170.77	0.04
CF_3CCl_3	87.04	86.97	−0.07
CF_2CCl_2	95.67	95.98	0.31
CF_3CCF_3	59.29	59.41	0.12
C_4F_8	140.85	140.84	−0.01
CF_4	—	68.70	—
$C_6H_5CF_3$	—	69.94	—
$CF_3CHClBr$	—	82.66	—
α-C_6F_{14}	86.33	86.22	−0.11
β-C_6F_{14}	130.37	130.26	−0.11
γ-C_6F_{14}	126.73	126.44	−0.29
Theoretical	0.19	0.05	0.19
errors	RET	REP	AET

[a] Reprinted with permission from E. R. Malinowski, *Anal. Chim. Acta,* **103,** 339 (1978).

The SPOIL, calculated to be 3.8, indicated a moderately acceptable test vector. This implied that the experimental error in the vapor shifts was approximately four times the error in the solution shifts, in agreement with the known experimental estimates that the error in the vapor shifts was 0.14 ppm, whereas the error in the solution shifts was 0.035 ppm.

Using the experimental estimate of the error in the test vector (0.14 ppm), the RELI was calculated to be approximately 80%. A RELI greater than 50% gives evidence that the test vector is a real factor.

Because the REP is smaller than the RET, the predicted gas-phase shifts are more accurate than the measured shifts. We see here, that, in certain situations, TFA can be used not only to identify a true factor but also to purify the target data.

4.7 ERRORS IN THE LOADINGS

In many studies the identities of the true factors may be known. Numerical values for the real row cofactors, which comprise the target vectors, may also be accurately known. Upon inserting these test vectors into the target-combination procedure, one is faced with the task of estimating the reliability of the loadings (i.e., the column cofactors, $\bar{c}_{jk}$), which result from the combination–transformation step. At the present time there exists two methods for estimating the errors in the loadings. One method[24] is based upon the theory of error for factor analysis and the second method[25] is based upon a statistical procedure called the jackknife method. The first method yields a root-mean-square error for each factor; the second method, although requiring considerably more computational time, yields the error and confidence limit for each individual loading. These two methods are discussed in the next two sections.

4.7.1 Estimation Based on the Theory of Error

The RMS of the error in the factor loading, EFL, can be deduced from the theory of error[2,23,24] previously discussed. If the errors are reasonably small, $r_{ij}^{*}/r_{ij} \simeq 1$ and, according to (4.21), the abstract column cofactor can be expressed as

$$c_{jk} = c_{jk}^{*} + \sigma_{jk}^{\ddagger} \tag{4.108}$$

which, in matrix form, can be written as

$$[C^{\ddagger}] = [C^{*}] + [E_{c}^{\ddagger}] \tag{4.109}$$

where $\sigma_{jk}^{\ddagger}$ is an element of the error matrix $[E_{c}^{\ddagger}]$ associated with the column-factor matrix.

The abstract column matrix $[C^{\ddagger}]$ can be transformed into the real matrix $[\bar{C}]$ by means of (3.15). Hence

$$[\bar{C}] = [\bar{C}^{*}] + [T]^{-1}[E_{c}^{\ddagger}] \tag{4.110}$$

where $[T]^{-1}[C^{*}] = [\bar{C}^{*}]$, assuming, of course, that $[T]$ contains little or no error. Since $[\bar{C}^{*}]$ is the real matrix, free from error, we find, from (4.110), that the error in the real loading matrix $[\bar{E}_{c}]$ is simply

$$[\bar{E}_{c}] = [\bar{C}] - [\bar{C}^{*}] = [T]^{-1}[E_{c}^{\ddagger}] \tag{4.111}$$

In order to obtain the RMS errors we consider the product $[\bar{E}_{c}][\bar{E}_{c}]^{T}$, which from (4.111) yields

$$[\bar{E}_{c}][\bar{E}_{c}]^{T} = [T]^{-1}[E_{c}^{\ddagger}][E_{c}^{\ddagger}]^{T}\{[T]^{-1}\}^{T} \tag{4.112}$$

According to the theory of error for AFA, $[E_{c}^{\ddagger}][E_{c}^{\ddagger}]^{T}$ is a diagonal matrix where the jth diagonal element equals $\sum_{k=1}^{k=c} (\sigma_{jk}^{\ddagger})^2 = c(\text{RE})^2/\lambda_j^{\ddagger}$. Thus

$$[E_c^{\ddagger}][E_c^{\ddagger}]^T = c(\text{RE})^2[\lambda^{\ddagger}]^{-1} \qquad (4.113)$$

where $[\lambda^{\ddagger}]^{-1}$ is a diagonal matrix composed of the reciprocals of the primary eigenvalues. Upon putting (4.113) into (4.112), we see that

$$[\overline{E}_c][\overline{E}_c]^T = c(\text{RE})^2[\tilde{T}][\tilde{T}]^T \qquad (4.114)$$

where

$$[\tilde{T}] = [T]^{-1}[\lambda^{\ddagger}]^{-1/2} \qquad (4.115)$$

Since the jth diagonal element of $[\overline{E}_c][\overline{E}_c]^T$ equals $c(\text{EFL})_j^2$, where $(\text{EFL})_j$ is the RMS of the error in the jth factor loading, we find that

$$(\text{EFL})_j = \text{RE}\sqrt{\tilde{T}_j \cdot \tilde{T}_j} \qquad (4.116)$$

where $\tilde{T}_j$ is the jth row of $[\tilde{T}]$ and $\tilde{T}_j \cdot \tilde{T}_j$ is the inner product.

Equation (4.116) provides an easy method for determining the RMS error in each factor loading, because RE is obtained from AFA and $\tilde{T}_j$ is obtained from TFA. An example calculation involving the use of this equation is given in Section 5.2.3.

4.7.2   Jackknife Method

Weiner and coworkers[25] adopted the *jackknife method* of Mosteller and Tukey[26] to obtain not only the error in each loading factor but also the confidence limits. The basis of the jackknife method is best understood by means of a simple example involving integers. Consider the following five numbers: 3, 5, 7, 10, and 15. Any one of these numbers (say, 7) can be expressed as the difference between the weighted mean of all the numbers and the mean formed by excluding the number in question. For example,

$$7 = 5\left(\frac{3 + 5 + 7 + 10 + 15}{5}\right) - 4\left(\frac{3 + 5 + 10 + 15}{4}\right) \qquad (4.117)$$

Although this calculation appears to be trivial for equally weighted integers, it is valuable when applied to other quantities, such as the weightings obtained from regression analysis, or the loadings obtained from target factor analysis. In the latter case, instead of integers, we deal with the overall loading $\overline{c}_{jk}(\text{all})$, obtained by using all the data, and the reduced loading $\overline{c}_{jk}(\text{red}, i)$ obtained when the data matrix is reduced by removing the ith row. From these we calculate the effective loading $\overline{c}_{jk}(\text{eff}, i)$ as follows:

$$\overline{c}_{jk}(\text{eff}, i) = r\overline{c}_{jk}(\text{all}) - (r - 1)\overline{c}_{jk}(\text{red}, i) \qquad (4.118)$$

where r corresponds to the number of rows in the complete data matrix. Values for $\bar{c}_{jk}(\text{eff}, i)$ are calculated by removing, systematically, each row of the data matrix, carrying out the combination target transformation each time, and applying (4.118).

Equation (4.118) is analogous to the integer calculation (4.117). The $\bar{c}_{jk}(\text{eff}, i)$ terms reflect the intrinsic effects of the data points that were dropped from the analysis. There are r values for each $\bar{c}_{jk}(\text{eff}, i)$. These values, according to Mosteller and Tukey, are normally distributed even though the original $\bar{c}_{jk}(\text{all})$ values may not be normally distributed. Approximate confidence limits may be calculated by applying standard statistical methods to the effective loadings.

The average of the effective loadings, $\bar{c}_{jk}(\text{av})$, represents the "best" estimate of the true loading.

$$\bar{c}_{jk}(\text{av}) = \frac{1}{r} \sum_{i=1}^{r} \bar{c}_{jk}(\text{eff}, i) \tag{4.119}$$

The variance, S_c^2, of the effective loadings can be calculated as follows:

$$S_c^2 = \frac{\sum_{i=1}^{r} [\bar{c}_{jk}(\text{eff}, i) - \bar{c}_{jk}(\text{av})]^2}{r - 1} \tag{4.120}$$

This allows us to estimate the confidence limits for each of the $n \times c$ loadings. For example, based upon the standard t-distribution, the confidence interval for $\bar{c}_{jk}(\text{av})$ is $\pm t S_c / \sqrt{r}$, where t is Student's coefficient for a given confidence level.

Weiner and coworkers[25] used the jackknife procedure to estimate the confidence limits of distribution constants obtained by TFA of GLC retention volumes. Theoretically, when both the surface areas and the volumes of the liquid phases are inserted into the TFA combination scheme as test vectors, the resulting loadings are the distribution constants: K_A, the adsorption constant at the gas–liquid interface, and K_L, the bulk liquid distribution constant, respectively. Confidence limits obtained from the jackknife method were compared to those obtained by treating the same data with multiple linear regression analysis. In all cases, the distribution constants determined by both procedures overlapped with respect to their stated confidence limits. In all cases, the confidence limits calculated by the jackknife method were larger than those calculated using the standard error equation of regression analysis. This is expected since the jackknife procedure takes into account errors associated with both controlled and uncontrolled variables, whereas the error criteria used in linear regression takes into consideration only those errors resulting from controlled variables. Of the two procedures, the jackknife method yields a truer picture of the confidence limits.

REFERENCES

1. E. R. Malinowski, in B. R. Kowalski (Ed.), *Chemometrics: Theory and Application,* ACS Symp. Ser. 52, American Chemical Society, Washington, D.C., 1977, Chap. 3.
2. E. R. Malinowski, *Anal. Chem.,* **49,** 606 (1977).
3. R. J. Rummel, *Applied Factor Analysis,* Northwestern University Press, Evanston, Ill., 1970.
4. H. H. Harman, *Modern Factor Analysis,* 3rd ed. Rev., University of Chicago Press, Chicago, 1967.
5. B. Fruchter, *Introduction to Factor Analysis,* D. Van Nostrand, Princeton, N.J., 1954.
6. A. L. Comrey, *A First Course in Factor Analysis,* Academic Press, New York, 1973.
7. J. T. Bulmer and H. F. Shurvell, *J. Phys. Chem.,* **77,** 256 (1973).
8. M. S. Bartlett, *Br. J. Psychol. Stat. Sect.,* **3,** 77 (1950).
9. Z. Z. Hugus, Jr., and A. A. El-Awady, *J. Phys. Chem.,* **75,** 2954 (1971).
10. J. J, Kankare, *Anal. Chem.,* **42,** 1322 (1970).
11. G. L. Ritter, S. R. Lowry, T. L. Isenhour, and C. L. Wilkins, *Anal. Chem.,* **48,** 591 (1976).
12. R. M. Wallace and S. M. Katz, *J. Phys. Chem.,* **68,** 3890 (1964).
13. D. L. Duewer, B. R. Kowalski, and J. L. Fasching, *Anal. Chem.,* **48,** 2002 (1976).
14. J. E. Davis, A. Shepard, N. Stanford, and L. B. Rodgers, *Anal. Chem.,* **46,** 821 (1974).
15. R. N. Cochran and F. H. Horne, *Anal. Chem.,* **49,** 846 (1977).
16. E. R. Malinowski, *Anal. Chem.,* **49,** 612 (1977).
17. R. B. Cattell, *Multivariate Behav. Res.,* **1,** 245 (1966).
18. R. W. Rozett and E. Petersen, *Anal. Chem.,* **47,** 1303 (1975).
19. H. F. Kaiser, *Educ. Psych. Meas.,* **20,** 141 (1960).
20. J. H. Kindsvater, P. H. Weiner, and T. J. Klingen, *Anal. Chem.,* **46,** 982 (1974).
21. O. Exner, *Collect. Czech. Chem. Commun.,* **31,** 3222 (1966).
22. R. B. Cattell, *Educ. Psych. Meas.,* **18,** 791 (1958).
23. E. R. Malinowski, *Anal. Chim. Acta,* **103,** 339 (1978).
24. E. R. Malinowski, unpublished work.
25. P. H. Weiner, H. L. Liao, and B. L. Karger, *Anal. Chem.,* **46,** 2182 (1974).
26. F. Mosteller and J. W. Tukey, in G. Lindzey and E. Aronson (Eds.), *The Handbook of Social Psychology,* 2nd ed., Vol. 2, Addison-Wesley, Reading, Mass., 1968, p. 134.

5 Numerical Examples

In order to illustrate the mathematical concepts discussed in Chapters 3 and 4, this chapter presents the complete details of a factor analysis calculation using a model set of data. Reference is made to the pertinent equations of earlier chapters so that the calculations at each step can be followed systematically. We use an example that can be solved with a hand calculator within a very short period of time so that those seriously interested in carrying out the calculations can do so without the aid of a computer. For this reason we restrict the calculation to a 3 × 3 matrix. Since it is necessary for us to devise a problem with fewer factors than rows or columns of the data matrix, we choose a two-factor space. This also allows us to make two-dimensional plots that can be used to study the transformations of the eigenvectors.

5.1   MODEL SET OF PURE DATA

With the foregoing restrictions in mind, we arbitrarily construct three linear equations of the type

$$d_{ik} = r_{i1}c_{1k} + r_{i2}c_{2k} \tag{5.1}$$

Numerical values for the column cofactors α, β, and γ and for the row cofactors $a, b, \ldots, j$ were arbitrarily chosen. These values are shown in matrix form in (5.2). Carrying out the matrix multiplication gives us the data matrix on the right-hand side of (5.2).

$$
\begin{array}{c}
\text{Row} \\
\text{Matrix}
\end{array}
\qquad\qquad\qquad\qquad\qquad
\begin{array}{c}
\text{Data Matrix}
\end{array}
$$

$$
\begin{array}{c}
\quad r_{i1} \quad r_{i2} \\
\begin{array}{c} a \\ b \\ c \\ d \\ e \\ f \\ g \\ h \\ i \\ j \end{array}
\begin{bmatrix}
0 & 4 \\
1 & -1 \\
2 & 0 \\
3 & 0 \\
4 & 3 \\
5 & -4 \\
6 & 5 \\
7 & 8 \\
8 & -2 \\
9 & -5
\end{bmatrix}
\end{array}
\begin{array}{c}
\\
\text{Column Matrix} \\
\quad \alpha \quad \beta \quad \gamma \\
\times
\begin{bmatrix}
2 & 5 & 2 \\
1 & 10 & -5
\end{bmatrix}
\begin{array}{c} c_{1k} \\ c_{2k} \end{array}
\end{array}
\;=\;
\begin{array}{c}
\quad\;\; \alpha \quad\;\; \beta \quad\;\; \gamma \\
\begin{array}{c} a \\ b \\ c \\ d \\ e \\ f \\ g \\ h \\ i \\ j \end{array}
\begin{bmatrix}
4 & 40 & -20 \\
1 & -5 & 7 \\
4 & 10 & 4 \\
6 & 15 & 6 \\
11 & 50 & -7 \\
6 & -15 & 30 \\
17 & 80 & -13 \\
22 & 115 & -26 \\
14 & 20 & 26 \\
13 & -5 & 43
\end{bmatrix}
\end{array}
\tag{5.2}
$$

Factor analysis attempts to solve the reverse of what was just done in generating the data matrix. Starting with the data matrix, we want to obtain the row and column matrices.

5.1.1   Constructing Covariance and Correlation Matrices

The covariance matrix, which we label $[Z]$, is calculated by taking the dot products of the column vectors of the data matrix shown in (5.2). In this case the normalization process is omitted. Carrying out the multiplication indicated by (3.17), we obtain

$$
[Z] = \begin{bmatrix}
1{,}364 & 4{,}850 & 212 \\
4{,}850 & 24{,}725 & -5{,}230 \\
212 & -5{,}230 & 4{,}820
\end{bmatrix}
\tag{5.3}
$$

Either the correlation matrix or the covariance matrix can be used in the factor analysis scheme. Both will yield results consistent with their own latent statistical criterion. We use the covariance matrix in our detailed example and thus regenerate the data matrix directly. If we choose to use the correlation matrix, we would, of course, regenerate the normalized data matrix instead. The normalized data can be converted into the original data by dividing by the appropriate normalization constants. The normalization constant, N_k, is determined by taking the square root of the reciprocal of the sum of the squares of each element in the kth column of the data matrix:

$$N_k = \left[\frac{1}{\sum_i d_{ik}^2} \right]^{1/2} \tag{5.4}$$

In this way we obtain the normalized data matrix, $[D]_N$:

$$[D]_N = \begin{bmatrix} 0.10831 & 0.25538 & -0.28807 \\ 0.02708 & -0.03180 & 0.10083 \\ 0.10831 & 0.06360 & 0.05762 \\ 0.16246 & 0.09539 & 0.08642 \\ 0.29784 & 0.31798 & -0.10083 \\ 0.16246 & -0.09539 & 0.43211 \\ 0.46029 & 0.50877 & -0.18725 \\ 0.59567 & 0.73135 & -0.37450 \\ 0.37906 & 0.12719 & 0.37450 \\ 0.35199 & -0.03180 & 0.61936 \end{bmatrix} \tag{5.5}$$

$$N_k \quad 2.7076 \times 10^{-2} \quad 6.3596 \times 10^{-3} \quad 1.44037 \times 10^{-2}$$

The respective normalization constants are listed below each column in (5.5).

The correlation matrix $[Z]_N$ is computed by taking dot products of the column vectors of the normalized data matrix in accord with (3.17). In this manner we obtain

$$[Z]_N = \begin{bmatrix} 1.00000 & 0.83513 & 0.08268 \\ 0.83513 & 1.00000 & -0.47908 \\ 0.08268 & -0.47908 & 1.00000 \end{bmatrix} \tag{5.6}$$

5.1.2 Decomposition of the Covariance Matrix

The next step is to determine the eigenvectors that span the space. A theorem of factor analysis states that the eigenvectors of the covariance matrix are the same vectors that span the space of the data matrix. To obtain the eigenvectors, we employ the method of iteration described in Sections 3.3.2 and 3.3.3.

To begin the iteration, we refer to (3.44).

$$[Z]C_1 = \lambda_1 C_1 \qquad (3.44), (5.7)$$

Here, C_1 is the first eigenvector and λ_1 is the corresponding eigenvalue. As a first approximation we arbitrarily set

$$C_1' = (0.57735 \quad 0.57735 \quad 0.57735)$$

a normalized vector. In accord with (5.7), we multiply this vector by the covariance matrix, giving the following:

$$\begin{bmatrix} 1,364 & 4,850 & 212 \\ 4,850 & 24,725 & -5,230 \\ 212 & -5,230 & 4,820 \end{bmatrix} \begin{bmatrix} 0.57735 \\ 0.57735 \\ 0.57735 \end{bmatrix} = \begin{bmatrix} 3,710.0 \\ 14,055.6 \\ -114.3 \end{bmatrix}$$

The resulting column vector on the right is then normalized by dividing each element by the square root of the sum of the squares of the elements:

$$\begin{bmatrix} 3,710.0 \\ 14,055.6 \\ -114.3 \end{bmatrix} = 14,537.4 \begin{bmatrix} 0.25520 \\ 0.96686 \\ -0.00008 \end{bmatrix}$$

The reciprocal of the normalization constant (14,537.4) is an approximation to λ_1. As a second approximation to C_1, we use the normalized vector

$$(0.25520 \quad 0.96686 \quad -0.00008)$$

resulting from the computation shown above. The new C_1 is multiplied by $[Z]$ and the resulting column vector is again normalized, giving a better approximation to C_1 and λ_1. This process is repeated again and again, each time generating newer and better approximations to C_1' and λ_1, until (5.7) is satisfied. Such iteration finally yields

$$\lambda_1 = 26,868.9$$

and

$$C_1' = (0.180200 \quad 0.957463 \quad -0.225372)$$

To obtain the second eigenvector, we proceed by calculating the first-residual matrix, $[\mathcal{R}]_1$, as dictated by (3.56):

$$[\mathcal{R}]_1 = [Z] - \lambda_1 C_1 C_1' \qquad (3.56), (5.8)$$

Our computations are as follows:

$$\lambda_1 C_1 C_1' = 26{,}868.9 \begin{bmatrix} 0.180200 \\ 0.957463 \\ -0.225372 \end{bmatrix} (0.180200 \quad 0.957463 \quad -0.225372)$$

$$= \begin{bmatrix} 872.49 & 4{,}635.82 & -1{,}091.21 \\ 4{,}635.82 & 24{,}631.67 & -5{,}797.91 \\ -1{,}091.21 & -5{,}797.91 & 1{,}364.74 \end{bmatrix}$$

Subtracting this matrix from the covariance matrix gives us the first-residual matrix:

$$[\mathscr{R}]_1 = \begin{bmatrix} 491.51 & 214.18 & 1{,}303.21 \\ 214.18 & 93.33 & 567.91 \\ 1{,}303.21 & 567.91 & 3{,}455.26 \end{bmatrix}$$

The second eigenvector is obtained by an iteration procedure analogous to the method used to obtain the first eigenvector, but involving (3.57):

$$[\mathscr{R}]_1 C_2 = \lambda_2 C_2 \tag{3.57), (5.9}$$

To start the iteration, we arbitrarily choose values for C_2, normalize C_2, and continue to apply (5.9) until the normalized elements of C_2 converge to constant values. In this way we find that

$$\lambda_2 = 4{,}040.2$$

and

$$C_2' = (0.348777 \quad 0.151999 \quad 0.924790)$$

The second-residual matrix is obtained by means of (3.59), which can be written as

$$[\mathscr{R}]_2 = [\mathscr{R}]_1 - \lambda_2 C_2 C_2' \tag{5.10}$$

Carrying out this computation, we obtain

$$[\mathscr{R}]_2 = \begin{bmatrix} 0.0 & 0.0 & 0.1 \\ 0.0 & 0.0 & 0.0 \\ 0.1 & 0.0 & 0.0 \end{bmatrix}$$

This residual is essentially zero. The small finite values are due to roundoff in the computations.

If correlation factor analysis had been used, the procedure would have been similar but the eigenvectors and eigenvalues would have been different because the computations would have involved $[Z]_N$ instead of $[Z]$.

5.1.3   Abstract Column and Row Matrices

Since two eigenvectors, C_1 and C_2, adequately account for the covariance matrix, we conclude that the data space is two-dimensional. This is in accord with the known facts. The abstract column matrix is constructed from the eigenvectors as discussed previously [see (3.71)], so that

$$[C^{\pm}] = \begin{bmatrix} C'_1 \\ C'_2 \end{bmatrix} = \begin{bmatrix} 0.180200 & 0.957463 & -0.225372 \\ 0.348777 & 0.151999 & 0.924790 \end{bmatrix} \tag{5.11}$$

According to (3.66), premultiplication of the inverse of the column matrix by the data matrix [given in (5.2)] yields the abstract row matrix. Because the eigenvectors are orthonormal, the inverse of the column matrix is simply its transpose. Carrying out this calculation, we find that

$$[R^{\pm}] = [R_1 \quad R_2] = \begin{bmatrix} 43.5267 & -11.0207 \\ -6.1847 & 6.0624 \\ 9.3939 & 6.6143 \\ 14.0909 & 9.9214 \\ 51.4330 & 4.9630 \\ -20.0419 & 27.5564 \\ 82.5903 & 6.0669 \\ 119.9323 & 1.1084 \\ 15.8124 & 31.9674 \\ -12.1357 & 43.5401 \end{bmatrix} \tag{5.12}$$

Multiplying, vectorially, any row of $[R]$ by any column of $[C]$ regenerates the original data value. For example, for row designee g and column designee β, we find that

$$d_{g\beta} = (82.5903)(0.957463) + (6.0669)(0.151999) = 80.000$$

This number is precisely the same as that found in the data matrix shown in (5.2).

5.1.4   Target Transformation

Although we have been able to express the data matrix as a product of two matrices, one associated with row designees and one associated with column

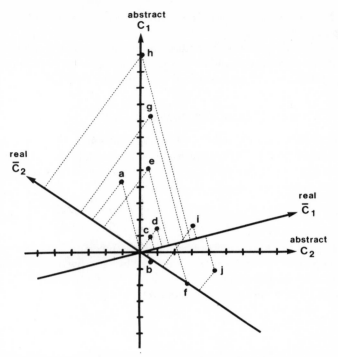

Fig. 5.1. Geometrical relationship between the abstract factor axes and the real axes for a two-dimensional example problem.

designees, the abstract row matrix (5.12) generated by factor analysis is quite different from the original row matrix (5.2). The reason for this difference lies in the fact that the factor analysis procedure selects a least-squares set of axes which span the factor space. At this point we have achieved the short-circuit reproduction shown in Figure 3.1.

Our next step is to find a new set of axes which allows us to transform the abstract row and column matrices into the original matrices, which have chemical meaning. The relationship between these two sets of axes is best viewed graphically. In Figure 5.1 the results of the factor analysis are plotted using the two row designee vectors C_1 and C_2 as the respective vertical and horizontal coordinate axes. To determine the score of a row designee on one of the new axes, we draw a line from the point perpendicular to the axis and read the relative distance along the axis. The scores (the projections) of the row-designee points onto the C_1 and C_2 axes give the respective values for the row cofactors listed in (5.12). Upon careful inspection, we can find two other axes, labeled $\overline{C}_1$ and $\overline{C}_2$, which yield projections correspondings to the original model set of row cofactors listed in (5.2). Thus by proper transformation of the coordinate axes, we are able to find new axes, having physical significance, which express the scores in accord with those given in (5.2).

For two-dimensional problems, graphical transformation is easily accomplished. For higher-order dimensions, graphical techniques can display only two-dimensional projections, making it virtually impossible to find the original basic axes. Numerical methods, such as the least-squares method described in Section 3.4, play a vital role in locating the fundamental axes. The least-squares target transformation vector required to locate such a fundamental axis is readily obtained by carrying out the calculation expressed by (3.97):

$$T_l = [\lambda^{\pm}]^{-1}[R^{\pm}]^T\overline{\overline{R}}_l \qquad (3.97), (5.13)$$

Here the first matrix on the right is a diagonal matrix containing the reciprocals of the eigenvalues. The second matrix is the transpose of the abstract row matrix. The third term, the test vector $\overline{\overline{R}}_l$, is a column of the original row matrix shown in (5.2), meant to simulate a real, basic factor.

In actual practice we test physically significant factors one at a time by using (3.84):

$$\overline{R}_l = [R^{\pm}]T_l \qquad . \qquad (3.84), (5.14)$$

If the suspected test vector $\overline{\overline{R}}_l$ is a real factor, the regeneration, according to (3.84), will be successful. In other words, each element of $\overline{R}_l$ will reasonably equal the corresponding element of $\overline{\overline{R}}_l$.

For our problem

$$[\lambda^{\pm}]^{-i} = \begin{bmatrix} 26{,}868.9 & 0.0 \\ 0.0 & 4{,}040.2 \end{bmatrix}^{-1} = \begin{bmatrix} 3.7218 \times 10^{-5} & 0.0 \\ 0.0 & 24.751 \times 10^{-5} \end{bmatrix} \qquad (5.15)$$

Inserting numerical values for $[\lambda^{\pm}]^{-1}$ and $[R^{\pm}]$ into (5.13), we find that

$$T_l = \begin{bmatrix} 0.0016200 & -0.0002302 & \cdots & -0.0004517 \\ -0.0027277 & 0.0015005 & \cdots & 0.0107767 \end{bmatrix} \overline{\overline{R}}_l \qquad (5.16)$$

5.1.5   Examples of Target Transformations

Next we give examples of two successful target transformations and one unsuccessful transformation. To test column r_{i1} of the row matrix given in (5.2), we let the elements of r_{i1} constitute $\overline{\overline{R}}_1$ and perform the multiplication expressed by (5.16), obtaining

Table 5.1 **Results of target tests**

Row Designee	$\overline{\overline{R}}_1$ Test	$\overline{\overline{R}}_1$ Predicted	$\overline{\overline{R}}_2$ Test	$\overline{R}_2$ Predicted	$\overline{\overline{R}}_{unity}$ Test	$\overline{R}_{unity}$ Predicted
		Successful Tests			Unsuccessful Test	
a	0	−0.0001	4	4.0003	1	0.1376
b	1	0.9999	−1	−1.0001	1	0.1215
c	2	1.9997	0	−0.0001	1	0.3119
d	3	2.9996	0	−0.0001	1	0.4678
e	4	3.9994	3	3.0001	1	0.7270
f	5	4.9995	−4	−4.0005	1	0.6421
g	6	5.9991	5	5.0001	1	1.1077
h	7	6.9988	8	8.0004	1	1.3668
i	8	7.9990	−2	−2.0005	1	1.1787
j	9	8.9990	−5	−5.0008	1	1.2315

$$T_1 = \begin{bmatrix} 0.0563012 \\ 0.2223760 \end{bmatrix} \tag{5.17}$$

We then carry out the calculation directed by (5.14), recalling that $[R^+]$ is the abstract row matrix given in (5.12). We obtain the results, labeled $\overline{\overline{R}}_1$, shown in Table 5.1. These results correspond exactly, except for errors due to computational roundoff, with column r_{i1} of (5.2), and therefore this test vector is a true factor of the space (as it indeed is, by design).

To test column r_{i2} of (5.2), we let the elements of r_{i2} constitute $\overline{\overline{R}}_2$ and repeat the foregoing procedure, obtaining first

$$T_2 = \begin{bmatrix} 0.0675947 \\ -0.0960138 \end{bmatrix} \tag{5.18}$$

and then the values labeled $\overline{R}_2$ shown in Table 5.1. Again, each value compares exactly, except for computational roundoff, with the corresponding value in (5.2). Hence we conclude that this vector is also a true factor of the space.

Only true test vectors will yield results compatible to the criteria discussed above. As an example of an unsuccessful transformation, consider the possibility that a factor of the space is the unity factor,

$$\overline{\overline{R}}_{unity} = (1 \quad 1 \quad 1 \quad 1 \quad 1 \quad 1 \quad 1 \quad 1 \quad 1 \quad 1)$$

which, in reality, is a test for a constant, a term independent of the row designee. Upon inserting this vector into (5.16), we find that

$$T_{unity} = \begin{bmatrix} 0.0111064 \\ 0.0313793 \end{bmatrix} \tag{5.19}$$

Carrying out the calculation expressed in (5.14), we obtain the results shown in Table 5.1. Comparing $\overline{\overline{R}}_{\text{unity}}$ with $\overline{R}_{\text{unity}}$, we conclude that this test vector is not a factor of the space. Target transformation affords us with a unique opportunity to test vectors one at a time, without our having to specify all the vectors simultaneously as required in other methods, such as regression analysis.

5.1.6 Example of the Combination Step

The complete transformation matrix $[T]$ is simply a combination of the transformation vectors:

$$[T] = [T_1 \cdots T_n] \tag{5.20}$$

For our problem, using the two successful transformation vectors, (5.17) and (5.18), we find that

$$[T] = \begin{bmatrix} 0.0563012 & 0.0675947 \\ 0.2223760 & -0.0960138 \end{bmatrix} \tag{5.21}$$

According to (3.75), postmultiplying the abstract row matrix $[R^+]$ by $[T]$ yields $[\overline{R}]$, the row matrix in the new coordinate system:

$$[\overline{R}] = [R^+][T] \tag{3.75}, (5.22)$$

When this computation is performed, we obtain matrix $[\overline{R}]$, which is identical to $[R]$ in (5.2), except for computational error.

To obtain the column matrix $[\overline{C}]$ in the new coordinate system, we apply (3.78):

$$[\overline{C}] = [T]^{-1}[C^+] \tag{3.78}, (5.23)$$

Matrix $[C^+]$ is composed of the eigenvectors determined previously [see (5.11)]. Using standard mathematical methods, we calculate the inverse of the transformation matrix:

$$[T]^{-1} = \begin{bmatrix} 4.6980 & 3.3075 \\ 10.8810 & -2.7545 \end{bmatrix} \tag{5.24}$$

Thus, using (5.23) and the values in (5.11) and (5.24), we find that

$$[\overline{C}] = \begin{bmatrix} 2.0002 & 5.0009 & 1.9999 \\ 1.0000 & 9.9995 & -4.9996 \end{bmatrix} \tag{5.25}$$

This matrix is identical to the original column matrix shown in (5.2).

In this manner we have completed the combination step illustrated in Fig. 3.1, reproducing the experimental data with a set of physically significant linear parameters. The factor analysis is now essentially complete.

5.1.7   Relationship between Factor Axes and Data Space

We now show the relationship between the factor axes and the data space. Factor space is described by the minimum number of axes necessary to account for the data space. For our example problem the factor space can be described by the two columns of the abstract row matrix obtained from the factor analysis. As discussed earlier, these two columns are mutually orthogonal and can be normalized by dividing each column by the square root of its respective eigenvalue. The normalized row matrix, $[R^+]_N$, for our problem is

$$[R^+]_N = \begin{bmatrix} 0.265539 & -0.173384 \\ -0.037730 & 0.095377 \\ 0.057308 & 0.104060 \\ 0.085963 & 0.156089 \\ 0.313772 & 0.078081 \\ -0.122268 & 0.433534 \\ 0.503850 & 0.095448 \\ 0.731659 & 0.017438 \\ 0.096465 & 0.502930 \\ -0.074035 & 0.684999 \end{bmatrix} \tag{5.26}$$

In Figure 5.2 we have designated the vertical and horizontal axes as the two normalized axes C_1 and C_2 and have plotted points corresponding to the various row designees from the values given in (5.26). Upon examining the normalized data matrix of (5.5), we see that three axes can be drawn on this two-dimensional graph, each one being associated with a column of the normalized data matrix. The perpendicular projections, indicated by the dashed lines, from the points to the data axes (labeled D_α, D_β, and D_γ) intersect the axes at values corresponding to the "scores" of the row designees on the axes. These scores are identical to the data values listed in the normalized data matrix. Figure 5.2 is analogous to Figure 3.3, but shows considerably more detail. We can see clearly the geometrical relationships between the data axes and the axes resulting from factor analysis. The data axes are not mutually perpendicular, nor do they coincide with the axes of the factor analysis. Any such correspondence would be a mathematical coincidence.

In Chapter 3 we learned that the correlation matrix is composed of dot products of the normalized data matrix. Each element of the correlation matrix represents the cosine of the angle between the respective data axes. Referring to (5.6), we conclude that the angle between D_α and D_β should be 33°22′, between D_α and D_γ should be 85°16′, and between D_β and D_γ should be 118°37′. In Figure 5.2 we see that this is true.

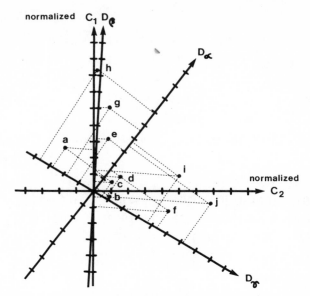

Fig. 5.2. Geometrical relationship between the normalized abstract factor axes and the normalized data columns. The length of each axis is 1 unit.

5.1.8   Predictions Based on Typical Vectors

There is no need for the factor analyst to identify the true, basic factors if he or she is simply interested in predicting new data. For example, the analyst may be primarily interested in predicting data values for new row designees, $k, l, m, n, \ldots$, which were not included in the original data matrix. If d_β and d_γ for the new row designees were known, we can predict d_α values using a key set of typical vectors as described in Sections 2.7 and 3.4.6. Typical vectors are simply columns of the data matrix. Since column vectors D_β and D_γ lie in the factor space, as shown in Figure 5.2, they may be used to define the two-dimensional factor space instead of the principal eigenvectors C_1 and C_2. Hence D_α can be expressed as a linear combination of D_β and D_γ.

To achieve this mathematically, we employ (3.104) and (3.105), which, for two data columns, β and γ, becomes

$$[T]_{\text{key}} = [T_\beta \quad T_\gamma] = [\lambda]^{-1}[R]^T[D_\beta \quad D_\gamma] \tag{5.27}$$

Numerical values for the product $[\lambda]^{-1}[R]^T$ were evaluated as indicated in (5.16). Using these results and the two data columns given in (5.2), we find that

$$[T]_{\text{key}} = \begin{bmatrix} 0.0016200 & -0.0002302 & \cdots & 0.0004517 \\ -0.0027277 & 0.0015005 & \cdots & 0.0107767 \end{bmatrix} \begin{bmatrix} 40 & -20 \\ -5 & 7 \\ \vdots & \vdots \\ -5 & 43 \end{bmatrix}$$

$$= \begin{bmatrix} 0.9575 & -0.2254 \\ 0.1517 & 0.9248 \end{bmatrix} \qquad (5.28)$$

According to (3.108), the transformed column factor matrix is calculated as follows:

$$[\overline{C}] = [T]_{\text{key}}^{-1}[C^+]$$

$$= \begin{bmatrix} 1.0056 & 0.2451 \\ -0.1649 & 1.0411 \end{bmatrix} \begin{bmatrix} 0.180200 & 0.9547463 & -0.225372 \\ 0.348777 & 0.151999 & 0.924790 \end{bmatrix}$$

$$= \begin{bmatrix} 0.2667 & 0.9973 & 0.0000 \\ 0.3334 & 0.0008 & 1.0000 \end{bmatrix} \qquad (5.29)$$

From (3.106) we see that

$$[D] = [D_\alpha \quad D_\beta \quad D_\gamma] = [D]_{\text{key}}[\overline{C}] = [D_\beta \quad D_\gamma][\overline{C}] \qquad (5.30)$$

Hence we find, using the first column of (5.29),

$$d_\alpha = 0.2667 d_\beta + 0.3334 d_\gamma \qquad (5.31)$$

We can verify that this equation yields correct predictions for d_α by inserting values for d_β and d_γ given in (5.2) for any row designee. For example, for row designee e, (5.31) predicts that

$$d_\alpha = 0.2667(50) + 0.3334(-7) = 11.00 \qquad (5.32)$$

which is in accord with the true value. Similarly, the d_α value for any new row designee can also be predicted from known values of d_β and d_γ, provided that the new row designee lies in the same factor space.

5.2 MODEL SET OF RAW DATA

The example studied in Section 5.1 is idealized because the data do not contain any error. Real data matrices inherently possess experimental uncertainties that tend to complicate the factor analysis. In this section we introduce artificial error into the data matrix and the target data. We then carry out the factor analysis and study some aspects of the theory of error described in Chapter 4.

5.2.1   Error in the Data Matrix

Our artificial error matrix, $[E]$, is the following:

$$[E] = \begin{bmatrix} -0.1 & 0.6 & 0 \\ 0.3 & 0.5 & 0.3 \\ -0.7 & 0.2 & -1.0 \\ 0.9 & 0.2 & -0.5 \\ 0 & -0.5 & 0.5 \\ 0.1 & -0.2 & 0.4 \\ -0.6 & 0.4 & -0.4 \\ 0.6 & -0.9 & -0.3 \\ -0.1 & -0.3 & 0.8 \\ 0 & 0.1 & 0 \end{bmatrix} \qquad (5.33)$$

These errors were generated by randomly selecting numbers between -1.0 and $+1.0$, rounding the nearest tenth unit. The root mean square of these errors is ± 0.477. When these errors are added, matrix-wise, to the pure data matrix of (5.2), we obtain the raw data matrix:

$$[D]_{\text{raw}} = \begin{bmatrix} 3.9 & 40.6 & -20.0 \\ 1.3 & -4.5 & 7.3 \\ 3.3 & 10.2 & 3.0 \\ 6.9 & 15.2 & 5.5 \\ 11.0 & 49.5 & -6.5 \\ 6.1 & -15.2 & 30.4 \\ 16.4 & 80.4 & -13.4 \\ 22.6 & 114.1 & -26.3 \\ 13.9 & 19.7 & 26.8 \\ 13.0 & -4.9 & 43.0 \end{bmatrix} \qquad (5.34)$$

This matrix is a better representation of real chemical data because it contains some uncertainty, known to be approximately ± 0.5.

From the raw data matrix, a somewhat different covariance (or correlation) matrix results. Carrying out the decomposition process as detailed in Section 5.1.2 leads to three eigenvectors instead of two as found for the pure data. The extra eigenvector is a secondary axis produced by the random error. The resulting eigenvalues, eigenvectors (grouped as the transposed, abstract column matrix), and the abstract row matrix are given in Table 5.2.

From the eigenvalues we can calculate the real error (RE), assuming that only only one or two eigenvalues belong to the primary set of eigenvectors. Referring to (4.44) we carry out the following computations:

Table 5.2   Results of factor-analyzing the raw data matrix using covariance about the origin

	Factor		
	1	2	3
Eigenvalue:	26,760.421	4,090.5369	2.0717
Abstract column matrix (transposed)	$\begin{bmatrix} 0.180847 \\ 0.956468 \\ -0.229048 \end{bmatrix}$	$\begin{matrix} 0.349120 \\ 0.155291 \\ 0.924121 \end{matrix}$	$\begin{matrix} 0.919462 \\ -0.247089 \\ -0.305838 \end{matrix}$
Abstract row matrix	$\begin{bmatrix} 44.1189 \\ -5.7411 \\ 9.6656 \\ 14.5264 \\ 50.8233 \\ -20.3982 \\ 82.9352 \\ 119.2441 \\ 15.2177 \\ -12.1847 \end{bmatrix}$	$\begin{matrix} -10.8161 \\ 6.5011 \\ 5.5084 \\ 9.8520 \\ 5.5204 \\ 27.8625 \\ 5.8277 \\ 1.3044 \\ 32.6784 \\ 43.5148 \end{matrix}$	$\begin{matrix} -0.3292 \\ 0.0746 \\ -0.4036 \\ 0.9064 \\ -0.1289 \\ 0.0670 \\ -0.6886 \\ 0.6305 \\ -0.2836 \\ 0.0127 \end{matrix}$
Real error (RE)	14.3049	0.4552	Undefined
Imbedded error (IE)	8.2589	0.3716	Undefined

1. Assuming that only the largest eigenvalue belongs to the primary set,

$$RE = \left[\frac{4{,}090.5369 + 2.0717}{10(3-1)}\right]^{1/2} = 14.3049 \qquad (5.35)$$

2. Assuming that the two largest eigenvalues belong to the primary set,

$$RE = \left[\frac{2.0717}{10(3-2)}\right]^{1/2} = 0.4552 \qquad (5.36)$$

The real error, 0.4552, calculated on the assumption that the two largest eigenvalues comprise the primary set, is in close agreement with the known root-mean-square error, 0.477. Hence the third factor belongs to the secondary set and contains nothing but error.

As discussed in Section 4.2, the removal of the secondary eigenvectors will lead to data improvement. Using (4.60), we can calculate the imbedded error (IE), the error that remains in the reproduced data matrix after deletion of the secondary eigenvectors. The results of such computations are shown in Table 5.2. By deleting the third eigenvector, we see that the error in the data is reduced from 0.455 to 0.372.

5.2.2   Error in the Target Data

In this section we study the effects of error on target transformation. First, we randomly select errors to be added to the two basic test vectors, r_{i1} and r_{i2}, utilized in Section 5.1.5. These errors, in a matrix form compatible with the data matrix, are:

$$\begin{bmatrix} 0.1 & -0.1 \\ -0.2 & 0 \\ 0 & 0.2 \\ -0.1 & -0.2 \\ 0.2 & 0.1 \\ 0.1 & -0.1 \\ -0.1 & 0 \\ 0 & 0.1 \\ -0.2 & 0.1 \\ 0.1 & 0.1 \end{bmatrix}$$

The RMSs of these two columns of errors are 0.130 and 0.118, respectively. When these two error vectors are added matrix-wise to the row matrix of (5.2), we obtain the two impure test vectors shown in Table 5.3, labeled "impure factor 1" and "impure factor 2." These two test vectors simulate real chemical data that contain experimental error.

When we subject these two impure targets to individual least-squares target transformations using the two principal factors of Table 5.2, we obtain the predicted vectors listed in Table 5.3. The appearent error in the target (AET) for each of these tests is obtained by calculating the root mean square of the differences between the test and predicted points, in accord with (4.88). The real error in the predicted vector (REP) is calculated by multiplying the RE in Table 5.2 by the lengths of the respective transformation vectors listed in Table 5.3. For example, according to (4.104), for impure factor one, we find that

$$REP = RE\sqrt{T \cdot T} = 0.4552[(0.05615)^2 + (0.22085)^2]^{1/2} = 0.1037 \quad (5.37)$$

Using the values obtained for AET and REP, we then calculate the real error in the target (RET) by means of (4.99). For both impure test vectors, the RET values, 0.137 and 0.113, are in excellent agreement with the known RMS errors in each vector, 0.130 and 0.118, respectively.

The SPOIL values, calculated by taking the ratio of the RET and REP values as defined by (4.107), are 1.32 and 2.13, respectively. Both of these values fall in the range of acceptability.

To calculate the RELI, we use (4.105). To represent $(RET)_{est}$ we employ the known RMS error in the target. For impure factor 1, the RELI is calculated to be 97%. For impure factor 2, the RELI is considered to be 100%, since $(RET)_{est}$

Table 5.3   Results of testing impure targets on the raw data matrix

Row Designee	Impure Factor 1 Test	Impure Factor 1 Predicted	Impure Factor 2 Test	Impure Factor 2 Predicted	Unity Test Test	Unity Test Predicted
a	0.1	0.088	3.9	4.027	1.0	0.154
b	0.8	1.113	−1.0	−1.000	1.0	0.139
c	2.0	1.759	0.2	0.146	1.0	0.280
d	2.9	2.991	−0.2	0.072	1.0	0.470
e	4.2	4.073	3.1	2.958	1.0	0.739
f	5.1	5.008	−4.1	−3.998	1.0	0.643
g	5.9	5.944	5.0	5.125	1.0	1.106
h	7.0	6.983	8.1	8.030	1.0	1.370
i	7.8	8.071	−1.9	−2.014	1.0	1.190
j	9.1	8.926	−4.9	−4.900	1.0	1.223
Least-squares	0.05615		0.06836		0.01114	
transformation vector	0.22085		−0.09346		0.03123	
AET	0.172		0.125		0.518	
REP	0.104		0.053		0.015	
RET	0.137		0.113		0.518	
Known RMS error in the target	0.130		0.118		(Unknown)	
SPOIL	1.32		2.16		34.3	
RELI	97%		100%[a]		(Unknown)	

[a] Since $(RET)_{est} > RET$, this test is considered to be 100% reliable.

is greater than the calculated RET. According to the SPOIL and RELI criteria, both of these test vectors are recognized to be true factors. When the unity test vector is targeted, the results (given in Table 5.3) clearly show that this vector is not acceptable because the SPOIL, 34.3, is excessively large. These tests demonstrate how we can identify real factors and reject those which are false.

5.2.3   Error in the Factor Loadings

Using impure factors 1 and 2 in combination, we obtain, by the procedure described in Section 5.1.6, the following transformed column matrix:

$$[C] = \begin{bmatrix} 2.0038 & 4.9155 & 2.0530 \\ 0.9996 & 9.9541 & -5.0368 \end{bmatrix} \tag{5.38}$$

These loadings contain some error as a consequence of the errors in the data matrix. A comparison of the foregoing values with the pure values given in (5.2) leads to the following error matrix:

$$[E_c^{\pm}] = [\overline{C}] - [\overline{C}*] = \begin{bmatrix} 0.0038 & -0.0845 & 0.0530 \\ -0.0004 & -0.0459 & 0.0368 \end{bmatrix} \qquad (5.39)$$

Hence the RMS errors associated with the first and second factor loadings are, respectively:

$$(EFL)_1 = \{[(0.0038)^2 + (-0.0845)^2 + (0.0530)^2] \div 3\}^{1/2} = \pm 0.058$$

and

$$(EFL)_2 = \{[(-0.0004)^2 + (-0.0459)^2 + (0.0368)^2] \div 3\}^{1/2} = \pm 0.034 \qquad (5.40)$$

With real chemical data we cannot perform such RMS calculations because we never know the values of the pure loadings, $[\overline{C}*]$. In order to estimate the RMS of the loading errors for each factor, we make use of (4.116):

$$(EFL)_j = RE\sqrt{\tilde{T}_j \cdot \tilde{T}_j} \qquad (4.116), (5.41)$$

where $\tilde{T}_j$ is a row of $[\tilde{T}]$, defined by (4.115):

$$[\tilde{T}] = [T]^{-1}[\lambda^{\pm}]^{1/2} \qquad (4.115), (5.42)$$

For the example problem:

$$\begin{aligned} [\tilde{T}] &= \begin{bmatrix} 0.05615 & 0.06836 \\ 0.22085 & -0.09346 \end{bmatrix}^{-1} \begin{bmatrix} (26{,}760.4)^{-1/2} & 0 \\ 0 & (4{,}090.5)^{-1/2} \end{bmatrix} \\ &= \begin{bmatrix} 0.02808 & 0.05254 \\ 0.06636 & -0.04315 \end{bmatrix} \end{aligned} \qquad (5.43)$$

Hence

$$(EFL)_1 = (0.4552)\{(0.02808)^2 + (0.05254)^2\}^{1/2} = \pm 0.027$$

and

$$(EFL)_2 = (0.4552)\{(0.06636)^2 + (-0.04315)^2\}^{1/2} = \pm 0.036 \qquad (5.44)$$

Considering that these calculations are based upon a statistically small number of data points, the results compare very favorably with those based upon a knowledge of $[\overline{C}*]$.

In the real world, problems are seldom as simple as this. In chemistry, data matrices usually require more than two dimensions. In spite of the complexity of the factor space, target factor analysis provides an efficient and realistic attack on such problems.

6 Procedures and Interpretations

6.1 SUMMARY OF PROCEDURES

This chapter contains practical instructions for planning, carrying out, and interpreting a factor analysis in chemistry. Procedures associated with each of the main steps outlined in Chapter 2 are discussed.

Factor analysis might involve as many as nine procedures: preparation, re-

118

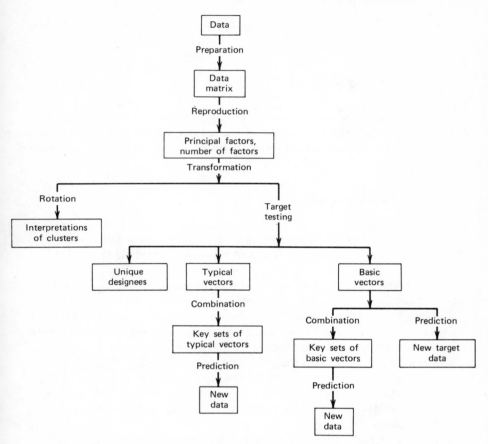

Fig. 6.1. Block diagram of the practical procedures in a complete factor analysis.

production, rotation, combinations of typical vectors, predictions using typical vectors, target testing to find unique designees, target testing to find basic vectors, combinations of basic vectors, and predictions using basic vectors. The interrelations between the procedures are shown in Figure 6.1.

Preparation and reproduction are required for all factor analyses. The first three procedures mentioned above involve abstract factor analysis, the middle three procedures involve routine applications of target factor analysis, and the last three procedures involve detailed uses of target factor analysis requiring greater chemical input. In many chemical problems, only a few of the procedures may be required to obtain the information desired. Chemists seeking the maximum insight into the nature of the factors will need to utilize the detailed target testing procedures, since the ultimate objective of factor analysis is to model data with basic vectors.

6.2   PREPARATION

The data preparation step involves three stages: problem selection, data selection, and data pretreatment. Chemists can anticipate frustration and eventual failure if the preparation procedures discussed in this section are carelessly carried out.

6.2.1   Problem Selection

Some general aspects of problem selection are discussed in this section. Thoughtful evaluation of the problem within the context of factor analysis is the first essential consideration in a well-planned research plan.

Factor Analyzability. One asks first and foremost whether the data might be modeled by a linear sum of product terms required for factor-analyzable solutions. The decision to apply factor analysis should be based, if possible, upon theoretical concepts or, if theory is lacking, upon a reasonable conceptual model for the data. Results of a factor analysis will be easier to evaluate if the researcher has prior intuition into the problem. Since the applicability of factor analysis may not be known a priori, factor analysis must often be initiated somewhat blindly. Fortunately, after the reproduction step has been carried out, the chemist can better judge whether to continue or to terminate the analysis.

Of the many kinds of data that might have factor analytical solutions, two kinds are especially suited for analysis. First, matrices involving spectral intensities at several wavenumbers for several multicomponent mixtures will usually have factor analytical solutions. As discussed in Section 1.3, each absorbance datum, at least for dilute solutions where Beer's law is obeyed, can be expressed as the sum of product terms consistent with the factor analytical model. Second, factor analytical solutions for solute–solvent problems are often anticipated intuitively if the observable is suspected to arise from a sum of chemical interactions involving solute–solvent interaction terms. Solute cofactors and solvent cofactors would then be associated with characteristics of individual molecules. For example, Hammett-like functions, based upon models consistent with factor analysis, explain large quantities of chemical data.

Type of Problem. Many kinds of data can be expressed as a matrix. Problems suited to factor analysis can be classified according to the types of designees represented by the row and column headings of the data matrix. Three types

Table 6.1   Examples of important types of data matrices

	Elements
Entity–property pair	
Molecule–physical property	Value of physical property
Molecule–spectral interval	Spectral intensity
Solution–spectral interval	Spectral intensity
Mixture–chromatogram interval	Chromatographic response
Entity–entity pair	
Solute–solvent	Spectral intensity
Solute–solvent	Chromatographic retention
Sample–chemical element	Analytical concentration

of designees—entities of matter, properties of matter, and events in time—can be used to describe chemical phenomenon. The word *entity* encompasses any sample of matter, from subatomic particle to stellar galaxy. Molecules, mixtures, instruments, and persons are examples of entities. *Properties* characterize and distinguish entities. Spectral interval, chromatographic interval, density, and temperature are examples of properties. *Events* specify the times at which measurements are made.

Based on the preceding classification of designees, we can form six types of data matrices: entity–entity, entity–property, entity–event, property–property, property–event, and event–event. For example, in an entity–property matrix, one set of designees is associated with a group of entities and the other set of designees is associated with a group of properties. In chemistry, entity–property, entity–entity, and possibly entity–event matrices are of particular importance. Entity–property matrices are by far the most studied type of matrix in the behavioral sciences. In the literature of abstract factor analysis, such matrices are called *R*-type matrices if the entities are the row designees and *Q*-type matrices if the properties are the row designees.

Examples of important entity–property and entity–entity matrices are given in Table 6.1. For each type of problem in the table, the nature of the row designee/column designee pair is furnished in the first column and the type of data factor-analyzed is furnished in the second column. The absorbance matrix in (1.11) is an entity–property matrix involving various mixtures (entities) at several wavenumbers (properties). A matrix of chromatographic data, involving two kinds of entities; solutes and solvents, forms an entity–entity matrix. An entity–entity matrix could be formed, for example, using the absorbances of several mixtures (entities) measured at several times (events). Many other examples of useful data matrices are discussed in Chapter 7 through 10.

Based on the data currently available in chemistry, a large number of entity–property matrices might be studied. For example, from Table 6.1 we see that both spectra and chromatograms of mixtures might be employed to form entity–property matrices. Such problems are of particular interest to analytical chemists because the number of components and the identity of the components in mixtures can be obtained from factor analysis. Entity–entity matrices, although less common, are of special interest to physical chemists because factor analysis can furnish insights into entity–entity interactions. Factor analyses of entity–event matrices offer new approaches to problems in chemical kinetics.

Information Desired. The kinds of information sought from a data analysis should be carefully considered before applying factor analysis. Factor analysis is just one of a host of chemometric methods,[1,2] each of which furnishes different kinds of information. Insights that might be obtained from factor analysis were summarized in Table 2.2. For comparison, the essential features of multiple regression analysis and pattern recognition are discussed in this section. These two approaches complement factor analysis more than any other chemometric method.

Multiple regression analysis[3] is a method for modeling data in which data (dependent variables) are expressed as a linear sum of a set of independent variables. The proportionality constant associated with each of the n independent variable in the regression model is called a *regression coefficient*. The best set of regression coefficients is calculated using a least-squares procedure. Regression coefficients are analogous to factor loadings.

Multiple regression analysis and target combination involve different approaches to complete data modeling. Multiple regression analysis has the practical advantage of furnishing force-fitted solutions having minimum error. However, combination TFA might be the preferred method for developing physically significant models for two reasons. First, the correct number of terms in a model can be obtained from the reproduction step of factor analysis, whereas in regression analysis the number of independent variables is chosen arbitrarily. Second, target testing provides a criterion for selecting the independent variables to be tested in combination, whereas in regression analysis independent variables must be selected more arbitrarily.

Pattern recognition[4] involves methods for classifying data. In chemistry, pattern recognition studies usually involve the calculation of a discriminant function which divides a "training set" of vectors into some specified binary classification. This function is then used to classify new vectors. Cluster analytical forms of pattern recognition are used to categorize vectors without the aid of a training set.

Pattern recognition and factor analysis are complementary techniques for analyzing matrices of data. The former approach is especially appropriate for classifying data, while the tour de force of TFA is factor interpretation. Factor analysis can also be employed for data classification in much the same manner as cluster analysis (see Section 6.7).

6.2.2   Data Selection

After selecting an appropriate problem, the next procedure is to assemble the data to be factor-analyzed. Data matrices for factor analysis should meet the four criteria discussed in this subsection. Careful data selection is a crucial prerequisite for successful factor analyses.

Reliable data should be chosen, especially since an accurate determination of the factor size usually hinges upon a knowledge of the error in the data. Although special methods have been developed to test the factor analyzability of the data matrix and to estimate the error in the data (see Section 4.4 and 4.5), there is no substitute for reliable data. Unless the chemist has a reasonable estimate of the error, factor analytical results cannot be interpreted with full confidence. If the data were collected in several laboratories, self-consistent data should be chosen and dubious data should be deleted. The uniqueness test (see Section 6.7.2) is useful for detecting points having gross errors.

For mathematical reasons, the data matrix must be complete. Lack of sufficient data will often prevent a chemist from carrying out factor analysis. However, although some points in a matrix may be missing, an analysis can still be conducted if smaller, but complete, submatrices can be formed. If the designees represented in the submatrices are selected carefully, useful factor analytical solutions can be obtained. Utilization of points estimated from statistical criteria should be avoided.

The size of the data matrix depends upon the availability of data, the objectives of the research, and the computational facilities at hand. To ensure general solutions, the largest possible matrix might be used. However, large matrices may be too complex to analyze initially. In order to make some progress, we may purposely work with a selected submatrix. Then, successively larger matrices can be analyzed to obtain more general solutions. If computational facilities are limited, matrices will have to be reduced to manageable size. Designees retained in submatrices should be chemically representative of the original sets of designees. When target tests are contemplated, the number of row designees and column designees should both preferably be at least twice the number of factors. This rule of thumb helps to ensure the mathematical soundness of target tests.

Designees in the matrix should adequately represent the problem the chemist wishes to study. If data for a number of molecules are available, the molecules chosen should cover the range of properties the researcher is interested in. Inclusion of too many kinds of designees may make the problem unnecessarily complicated and possibly block out the very properties one most wishes to study.

For instance, molecules with unique chemical properties introduce additional factors. Such molecules should be included only if the chemist is particularly interested in those molecules.

6.2.3   Data Pretreatment

There are three aspects of data pretreatment: (1) choosing covariance or correlation factor analysis, (2) equalizing the error in the matrix, and (3) selecting the functional form for the data. As much as possible, data pretreatment should be based on sound theoretical criteria. Uncritical pretreatment can severely hamper the interpretation of results and might even invalidate the factor analysis.

The decision to factor-analyze the raw data (covariance) or to normalize the data before factor analysis (correlation) is based upon the type of error in the data. As discussed in Section 3.2.1, covariance is used when the absolute errors of each data column are similar, and correlation is used when the relative (percent) errors of each data column are similar. Whenever a chemist is uncertain about the type of error in a problem, covariance is the preferred method because the majority of chemical measurements involve absolute, rather than relative, error. The two approaches lead to equivalent results only if the data points are similar in magnitude and have uniform errors.

Theoretically, data matrices in which the errors are uniform throughout are best suited for factor analysis. If errors of differing magnitudes are scattered throughout the matrix, smaller matrices with more uniform error should be formed. If this is not possible, the chi-squared criterion (see Section 4.3.1) should be used to determine the factor size. If the error varies considerably from column to column in the matrix yet is relatively constant within each column, "standardization" of the columns is recommended. Each column is divided by its standard deviation, thereby unitizing the error over the matrix. For example, if the four columns in a matrix have fairly constant absolute errors of 1, 4, 10, and 5, respectively, each element of the respective columns should be divided by 1, 4, 10, and 5, and covariance should be applied. Standardization should always be applied when the columns in the matrix involve different units.

Unless theory or empiricism suggests the most appropriate functional form for the data, factor analysis of the raw data and perhaps of the logarithm of the data should be carried out. Analyses utilizing other functional forms ordinarily serve no useful purpose. Lacking other criteria, chemists should use the functional form that will allow comparison of the factor analytic results with previous research in the field. Logarithmic transformation of the data may be particularly useful since logarithmic relations are quite common in chemistry. Empirical equations such as the Hammett functions suggest that the logarithm of equilibrium constants and of rate constants, rather than the raw constants, should be factor-analyzed.

6.3 REPRODUCTION

Reproduction involves calculating the principal factor solution and determining the correct number of factors. Since all subsequent procedures of factor analysis are based on the PFA model, the reproduction step should be conducted with exceptional care.

Principal factor eigenanalysis can be carried out routinely with many readily available computer programs.[5,6] Determining the rank of the data matrix, a not-so-routine procedure, entails, as described in Chapter 4, the use of both experimental error criteria and the theory of error for abstract factor analysis. Several examples of the determination of the number of factors were presented in Chapter 4. These examples illustrate the strategy employed in the reproduction step and should be reviewed carefully.

Once the number of factors is determined, the factor analyzability of the data can be established. Factor analysis is warranted if the number of factors is in line with the number expected from chemical reasoning, or, lacking insight into the complexity of the data, if considerable factor compression is indicated. For example, if a matrix having 10 columns and more than 10 rows requires four factors for adequate reproduction, the factor analysis can be continued with confidence since factor compression is considerable. However, if reproduction indicates nine factors, there is insignificant factor compression and the problem probably does not have a factor analytical solution. Such results suggest strongly that the problem be dropped. Blind, unjustified factor analyses should be viewed with skepticism.

If data are to be modeled accurately and particularly if target testing is to be applied, the true number of factors must be employed in the principal factor solution. If too few factors are used, important factors needed to describe the data accurately will be omitted from the analysis. Oversimplified solutions involving two or three factors seldom reproduce chemical data near experimental error. Using too many factors reproduces experimental error and leads to a frustrating search for extra factors which have no physical meaning. For target testing, an oversized factor space is particularly misleading because target tests will appear successful if the factor size is greatly overestimated.

Regardless of the true number of factors, a researcher may elect to limit the study to an arbitrary, small number of factors. Such results, although less accurate, are easier to interpret since only major classification features of the data are considered. Because the first two principal factors account for a major fraction of the variation in data in most chemical problems, two-dimensional plots are frequently employed to show dominant clusterings in the data. In such plots, each designee is represented by a point, the coordinates of each point being the cofactors for the specified designee on the first and second principal factors. Designees that lie near each other in this two-factor representation of the data are said to "cluster." Designees that lie far from all the other designees are said to be "unique." Examples from chemical problems are shown in Figures 9.1, 10.2, and 10.5. Similar plots based on rotated factors are even more informative.

6.4   ROTATION

Rotation involves choosing the type of rotation, performing the calculation, and interpreting the results. Chemists desiring detailed knowledge of rotational methodologies should consult monographs on abstract factor analysis written for behavioral scientists. We recommend in particular the treatise of Rummel.[7] A complete bibliography for abstract factor analysis follows Chapter 10.

Several types of rotations are incorporated in most of the statistical packages of computer programs. The *Statistical Analysis System* (SAS)[5] and the *Statistical Package for Social Sciences* (SPSS),[6] for example, are useful for carrying out rotations. Three orthogonal methods (varimax, quartimax, and equimax) and an oblique rotation (oblimin) are available in the standard packages. The theory behind these rotations is discussed in Section 3.4.2. Although varimax is the most widely used rotation, oblique methods may be of greater value for classifying chemical data.

Interpretation of rotated solutions entails a careful examination of the patterns in the rotated matrix. Two types of information, clusterings among designees and physical explanations for the factors, are sought. Cataloging information based on clusters of data is quite valuable to chemists. Similar designees have similar cofactors on a given factor, indicating the formation of a cluster in factor space. Many methods for comparing vectors mathematically are used to evaluate clusterings of designees, as described in Chapter 20 of the book by Rummel.[7] Unique designees, which exist separately in isolated regions of factor space, have cofactors that are different from any of the other designees. If a designee is the only one to exhibit a large cofactor on a given factor, the designee is uniquely associated with that factor. Plots of rotated cofactors involving two or three factors show the major clusterings of data. An example of a three-dimensional plot utilizing varimax rotation is given in Figure 10.1.

To develop physical explanations for rotated vectors, the factor analyst searches for real vectors that are similar to rotated vectors. In the behavioral sciences much effort is exerted trying to identify each rotated vector with hypothetical or experimental parameters (see Chapter 21 in the book by Rummel[7]). However, chemist must be quite cautious when attempting to interpret rotated factors physically. To expect a direct correspondence between a rotated vector and a basic factor is naive. Even less physical significance should be attached to principal factors, given the global-averaging nature of principal component analysis. Target testing, rather than rotation, should be employed to identify real factors.

6.5 COMBINATIONS OF TYPICAL VECTORS

Using the combination step, factor analytical solutions can be expressed efficiently as a small number of typical rows or typical columns from the data matrix (see Section 2.7). Finding key sets of typical vectors entails taking combinations of (1) columns from the original data matrix to characterize the key column designees, and (2) rows from the data matrix to characterize the key row designees.

The target factor analytical computer program[8] enables a chemist to test either selected combinations of typical vectors or all possible combinations of vectors from the original data matrix. The former approach may be sufficient if the chemist wishes to test only a few ideas or if computational capabilities are limited. The latter approach guarantees that the best set of typical vectors will be found, albeit at the expense of large amounts of computer time. The size of the data matrix and the number of vectors per combination determine the total number of combinations. For example, if the combination step involves a 12×14 matrix having a four-factor solution, the total number of combinations of typical columns is $14 \times 13 \times 12 \times 11/4! = 1001$. For large matrices, computer execution times for a complete combination step can easily exceed 1 hour. Researchers having smaller computers will necessarily have to run combinations on a reduced scale. An abbreviated combination step should include typical vectors which are of greatest interest to the chemist, which span a wide range of properties, and which have large uniqueness values (see Section 6.7.2).

Key typical vectors can be identified efficiently by the following elimination procedure. First, a small number of combinations selected from chemical insight is run to determine a preliminary, key set. Then other vectors are substituted one at a time for one of the vectors in the first set to find an improved, second key set. The substitution process is repeated as many times as desired, using at each stage the best set from the previous stage. In this manner, a useful solution may be generated from a relatively small number of combinations.

Several kinds of information are obtained using typical vectors in combination. The success of each combination model is judged by comparing the RMS error for a given model to the RMS experimental error. Quite good solutions of typical vectors are expected in chemical problems that have clear-cut factor analytical solutions. Generally, the RMS error for the key combination set will be only slightly greater than the RMS error for the abstract reproduction utilizing the same number of factors. The factors can be well represented with a small set of typical vectors. Results can be summarized, for a selected number of better combinations, by tabulating the designees involved and the RMS error. Designees represented most often in the best combinations can be compared to those selected from chemical insight alone.

Results from the combination step are useful for classifying designees. For example, similar designees can be identified in the following way. If two different combinations lead to nearly equal RMS errors, the two sets of vectors are equivalent. More specifically, if the substitution of one typical vector by another typical vector does not produce a significant change in the RMS error in combination testing, the two vectors that were interchanged are equivalent. The two designees belong to the same cluster. By comparing equivalencies from many combinations, an overall pattern of clusterings may be detected. A tabulation of the fraction of times each vector occurs in the better combinations is valuable for identifying the most important typical vectors. Vectors appearing most frequently in the best models are particularly important. An example of a fractional-representation table is discussed in Section 9.6.

6.6 PREDICTIONS FROM KEY SETS OF TYPICAL VECTORS

New data rows and new data columns for a data matrix can be predicted using key sets of typical vectors obtained from target combination.

To demonstrate the calculations, consider a 4×5 data matrix that obeys a two-factor model. We wish to predict complete rows of data for some new row designees, labeled $a, b, \ldots, m$. First, the pair of typical columns from the data matrix which gives the best data reproduction is found using target combination. Suppose that the key set in this example involves columns 2 and 3 from the data matrix. We then output the corresponding transformed column matrix:

$$
\begin{array}{lccccc}
 & \multicolumn{5}{c}{\text{column designee}} \\
\text{key factor} & 1 & 2 & 3 & 4 & 5 \\
\text{2 (associated with key column 2)} & c_{21} & 1 & 0 & c_{24} & c_{25} \\
\text{3 (associated with key column 3)} & c_{31} & 0 & 1 & c_{34} & c_{35}
\end{array}
\qquad (6.1)
$$

Here the cofactors c_{jk} are calculated from a combination using key columns 2 and 3. Next we form a new row matrix involving the new row designees:

$$
\begin{array}{cc}
\text{new row} & \text{key factor} \\
\text{designee} & 2 \quad 3 \\
a & \\
b & \\
\vdots & \\
m &
\end{array}
\begin{bmatrix}
d_{a2} & d_{a3} \\
d_{b2} & d_{b3} \\
\vdots & \vdots \\
d_{m2} & d_{m3}
\end{bmatrix}
\qquad (6.2)
$$

Here the cofactors d_{ij} are key data relating the new row designees and the key column designees. For any given row designee, values for both key cofactors d_{ij} must be known.

Premultiplying column matrix (6.1) by row matrix (6.2), we can generate a complete matrix of TFA-predicted data. For example, d_{b4}, which relates new row designee b and non-key-column designee 4, is obtained by multiplying the bth row of matrix (6.2) by the fourth column of matrix (6.1) term by term:

$$d_{b2}c_{24} + d_{b3}c_{34} = d_{b4}(\text{predicted}) \tag{6.3}$$

Equation (6.3) is of the same form as the basic equation of factor analysis.

The approach outlined in this section is a generalized version of the method detailed in Section 5.1.8.

Since key sets of typical vectors can be identified without any prior insight into the data, this mass production, target prediction procedure can be carried out routinely. Predictions based on typical vectors are expected to be reliable, assuming that the key set is a good model for the original data and provided that the new designees obey the same factor model as the original designees. The validity of predictions can be verified if experimental values are available for some of the predicted points.

6.7   TARGET   TESTING

If individual, basic factors are to be identified, target transformation procedures[8] should be used. Target testing requires more scientific input than any of the steps discussed previously in this chapter. Such transformations are of greatest value when based upon scientifically sound ideas. Indeed, without a prespecified model for the data, target factor analysis will be difficult and can degenerate into a hunting game. Theory, empirical insight, and intuition are employed to design test vectors and to interpret the results of target tests.

6.7.1   General Considerations

Two general requirements concerning the number of points needed for a test vector and the range spanned by the test values should be considered before carrying out a target test.

The number of points in a test vector must be greater than the number of factors employed in the transformation. We call this important but easy-to-overlook criterion the *greater-than-n rule*. To illustrate the rule, if a target test involves four factors, at least five test points should be incorporated in the test vector. Results become more reliable as more points are entered in the test vector. A test utilizing fewer than four points would be invalid and should not be run.

If four points are employed, the vector will be predicted perfectly. Such superficially successful results, which arise from a mathematical artifact of target factor analysis, are misleading and should be ignored.

As long as the greater-than-n rule is obeyed, test points can safely be omitted from the test vector. This freedom to "free float" is exceptionally useful because, in most tests, many of the test values may not be available. Very few chemical parameters could be tested without the use of the free-floating option. Furthermore, the success of a target test can be confirmed by free floating a few known values in a new test vector. If these deliberately deleted points are predicted adequately, the validity of the original test is strengthened.

The second general requirement for target tests concerns the range of values spanned by the test points. Each target vector should span, as completely as possibly, the entire range of values for the property being tested. Poorly defined test vectors that do not include representative high- and low-valued test points are likely to give different results from a well-defined test vector having a complete range of test values. In practice, we enter all the extremums available, hoping that the vector will span the complete range of the real factor. Interpolation is safer than extrapolation.

If there is much uncertainty in determining the number of factors, target tests should be carried out for $n - 1$, n, and $n + 1$ factors. As results accumulate, a firmer estimate of the number of factors might be made. For example, should several basic test factors transform poorly with n factors but successfully with $n + 1$ factors, the factor size may be one unit greater than was originally supposed. In some tests, values predicted for free-floated points may become less accurate if the number of factors is greater than the correct number. In such situations, the factor size may be one unit less than that originally supposed.

To obtain empirical correlations, various functional forms of test vectors, such as the square, reciprocal, and logarithm of test points, can be target tested. By testing different functional forms of test vectors, solutions with lower RMS errors will be obtained in target combination. However, as more functional forms are tested, the chemist becomes mired in a mathematical procedure lacking physical meaning.

6.7.2   Uniqueness Tests

A special kind of target test called the *uniqueness test* should be carried out routinely for both row and column designees in every TFA study. The uniqueness test is designed to identify designees behaving atypically due to either chemically unique properties or to gross errors in the data.

A uniqueness test vector contains an input value of 1 for the designee being tested for uniqueness and 0's for all the other designees. Uniqueness tests for each of the row designees and for each of the column designees should be per-

Table 6.2 Example of uniqueness test

| Row
Designee | Uniqueness Test
for Designee 2 | |
	Test Vector	Predicted Vector
1	0.00	0.37
2	1.00	0.85
3	0.00	0.16
4	0.00	0.18
5	0.00	0.20
6	0.00	0.42

formed. Qualitatively, a designee is considered unique if the "uniqueness" value predicted for that designee on its uniqueness test is large relative to all the other values on the predicted vector. The uniqueness test also serves as a preliminary classifier in that similar designees tend to have similar predicted values on uniqueness tests. Hence the predicted uniqueness vector may indicate clusterings of designees that have a factor in common.

Uniqueness tests can pinpoint designees associated with unsuspected gross errors. An "outlier" point creates a separate, unique factor and may increase the factor size by as much as one unit. Thus large uniqueness values may indicate unreliable data. Unless a designee is considered to be chemically unique, high uniqueness values are adequate grounds for removing the associated row or column from the data matrix, and then redoing the factor analysis on the reduced matrix. Removal of the suspect designee may result in better reproduction and in some problems may reduce the rank of the data matrix by one unit.

Results for a hypothetical uniqueness test are presented in Table 6.2. The row designees are specified by number in the left-hand column. For each row designee, the test value (input) is listed under the column titled "Test Vector," and the least-squares prediction (output) from target factor analysis is listed under "Predicted Vector." Row designee 2 is tested for uniqueness in this example. The predicted uniqueness value of 0.85 is sufficiently large that designee 2 is considered unique. The intermediate predicted values of 0.37 and 0.42 for designees 1 and 6, respectively, give a preliminary indication that these two designees cluster. A uniqueness test from a linear free-energy problem is shown in Table 10.3.

Results from uniqueness tests are usually tabulated, giving the uniqueness values for each designee on its test and listing the other designees on each test which have predicted values greater than say 0.15. The latter information is used to group similar designees into clusters. A uniqueness summary for a chromatographic problem is given in Table 9.5. Tabulation of uniqueness values for each designee over a range of factors may be valuable for linking unique behavior to particular factors. Thus if the uniqueness value for a certain designee increases dramatically when the factor size is increased from $j - 1$ to j, some unique property of that designee may account for the jth factor.

6.7.3   Unity Test

Another special test that should be run routinely is the *unity test*. This test vector, consisting entirely of 1's, tests for a constant factor common to all row designees. Unity is identified as a factor if the predicted values from target transformation are all near unity. Unity is expected to transform well if all the row designees incorporate the same functional group. If the first principal factor is particularly important, unity will test well regardless of the physical situation. Results for unity tests from four chemical problems are shown in Tables 9.8, 9.9, 10.2, and 10.3.

6.7.4   Designing Test Vectors

Designing chemically meaningful test vectors is the most important and the most difficult task in target factor analysis.

Test vectors can be based on theory, empiricism, or chemical insight. If theoretical or empirical models are available, the variables in the models dictate the form of an initial set of parameters to be target-tested. For example, in the absorbance problem posed in Section 1.3, Beer's law tells us that the test vectors should be related to the concentrations and to the molar absorptivities of the components. Whenever theoretical insight is lacking, an empirical approach involving chemical intuition must be employed. For example, because there is no general model for solute–solvent interactions, test vectors for chromatographic problems must be formed intuitively from chemical criteria.

When developing test vectors, chemists should apply not only theoretical and empirical knowledge of the problem being studied, but also broader knowledge obtained from related problems and from other, seemingly unrelated fields of chemistry. Basic vectors developed in one TFA study may be equally useful in different problems involving similar kinds of designees, espeically if the two problems have similar theoretical bases and therefore might have factors in common. For example, parameters tested originally in the target factor analysis of chromatographic retention data are expected to be reasonable targets in studies of solubility data and of other data having to do with solute–solvent interactions.

Test vectors of molecular entities are of two types:

1. Physical.
2. Structural.

Physical vectors are related to bulk physical properties of molecules, whereas structural vectors are based on detailed, structural properties of molecules. Enthalpy of formation, molar refraction, and boiling point are examples of physical properties. Handbooks, treatises, and review articles are important sources of physical-property test data. Carbon number, number of rings, and stretching frequency are examples of structural test parameters. Structural vectors are suggested not only from journal articles but also from chemical insight (see below). With both physical and structural parameters, sufficient data seldom are available to form a complete test vector. The privilege of free-floating missing test points, as long as the number of test points is greater than the number of factors, is of inestimable value in target factor analysis. In practice, scarcity of data or lack of insight can be a major hindrance to the development of usable test vectors.

Structural test vectors play a particularly valuable role in target factor analysis. Using structural vectors, chemists can test for factors that arise from differences in the structures of the entities. Theory, empiricism, and intuition all play parts in designing structural vectors. We create a structural vector by assigning a numerical scale to express some structural variation among the entities. Even quite subtle differences in structure can be target-tested. Many examples of structural vectors as well as physical vectors are given in Chapters 7 through 10.

The four hypothetical test vectors in Table 6.3 illustrate different types of structural vectors. The rationale behind each vector is stated at the bottom of the table. Test vector 1 in the table, exemplifying the simplest of structural vectors, is an extended form of the uniqueness test described in Section 6.7.2. Such "clustering" uniqueness tests contain 1's for designees believed to have the property being tested, 0's for designees believed to lack the tested property, and free-floated values for questionable designees. These "yes–no" binary classification vectors are useful, for example, to determine whether a specific functional group is responsible for a factor. Test vector 2 in the table utilizes a semiquantitative scale to rank the designees of the test vector. Such scales conveniently employ integral values for designees thought to have the property, 0's for designees lacking the property, and free-floated test points for uncertain or missing values.

Test vectors 3 and 4 in Table 6.3 are examples of quantitative vectors based on detailed theoretical or empirical knowledge. If two vectors are proportional to each other, the vector known with the greatest accuracy probably better represents the real factor. Test vectors 1 and 4 have quite similar patterns and

Table 6.3 Examples of structural test vectors

| Row | Test Vector | | | |
Designee	1	2	3	4
1	1	0	2.3	22.7
2	1	0	1.6	22.4
3	0	2	—	0.3
4	—	2	5.7	—
5	0	1	3.4	0.1
6	0	1	0.8	0.5

Idea being tested:

Test vector 1—Qualitative scale on which designees 1 and 2 exhibit the property, designees 3, 5, and 6 lack the property, and designee 4 is free-floated.

Test vector 2—Semiquantitative scale on which the property is weighted twice as heavily for designees 3 and 4 as for designees 5 and 6, and designees 1 and 2 lack the property.

Test vector 3—Quantitative scale on which designee 3 is free-floated.

Test vector 4—Quantitative scale on which designee 4 is free-floated.

are essentially equivalent to each other from a target factor analysis viewpoint. Realizing that test vector 1 is a simplified qualitative scale whereas test vector 4 is expressed to three significant figures, we expect that test vector 4 will be a better representation of a real factor.

Additional structural vectors may be suggested by noting which designees cluster on uniqueness tests and on rotated factors, and by determining which designees are equivalent in the combination of typical vectors (see Section 6.5). Test vectors that transform with borderline success may furnish valuable hints for the formation of new test vectors. For example, poorly predicted points on a test vector might be free-floated in the test vector. However, this kind of search for better and better test vectors should not deteriorate into a mathematical exercise devoid of chemical input.

6.7.5 Interpretations

Two complementary approaches, one based on mathematical criteria and the other based on qualitative comparisons, are employed in the difficult task of evaluating the results of a target test. The quantitative criteria have not yet been fully evaluated and the qualitative approach can be misleading since experimental error influences target testing in complicated ways. In practice, final evaluation should be based upon a combination of conclusions from both approaches.

One mathematical approach,[9] based on the effects of experimental error on target tests (see Section 4.6), appears to be particularly useful. A major advantage of this method is that the effect of both the error in the test points and the error in the data matrix are accounted for. Two functions, SPOIL and RELI (reliability), are used to evaluate target tests. Examples of the application of these two functions, given in Sections 4.6 and 5.2.2, should be studied carefully.

Many other mathematical and statistical criterion for estimating the similarity of two vectors are available. In Chapter 20 of his monograph, Rummel[7] concluded that the RMS coefficient, the congruence coefficient, and the intraclass correlation coefficient are especially useful for comparing vectors. Cattell[10] advocated the use of the congruence coefficient, a salient variable similarity index, configurational correlations, and the intraclass correlation coefficient for comparisons. The monographs should be referred to for definitions of these various quantities. For target tests, Howery and coworkers[11] recommended the intraclass correlation coefficient, while Rozett and Petersen[12] employed the relative factor error. Since test vectors seldom obey parametric statistics, statistical criterion developed for normally distributed data, such as the t-test and the F-test, are inappropriate for target factor analysis. Comparative studies of the various mathematical criteria within the context of target testing are needed.

To qualitatively evaluate target tests, the test vector and its corresponding predicted vectors are carefully compared point by point using chemical insight.

Table 6.4 Hypothetical results from target testing a complete test vector

Row Designee	Test Vector	Hypothetical Predicted Vector				
		1	2	3	4	5
1	283	281	292	244	279	147
2	134	138	131	111	230	273
3	42	44	59	73	32	114
4	187	180	177	235	188	215
5	255	258	250	216	261	122
6	91	85	83	123	94	156

Qualitative conclusions (assuming two factors, error in test points less than 5):

Result 1—Very good similarity; test vector is a basic factor.

Result 2—Good similarity; test vector is a basic factor.

Result 3—Fair similarity; test vector is a borderline basic factor.

Result 4—Good similarity, except for poorly predicted value for designee 2; test vector is not a basic factor.

Result 5—Poor similarity; test vector is not a basic factor.

Table 6.5   **Hypothetical results from target testing an incomplete test vector**

Row Designee	Test Vector	Hypothetical Predicted Vector			
		1	2	3	4
1	114	119	132	55	97
2	63	66	56	79	83
3	—	98	94	112	24
4	—	57	61	−2	−36
5	88	80	81	103	73
6	—	73	103	94	179

Qualitative conclusions (assuming two factors, error in test points less than 5):

Result 1—Very good similarity; test vector is a basic factor; predicted values for free-floated points probably are trustworthy.

Result 2—Fair similarity; test vector is a borderline basic factor; predicted values for free-floated points may be trustworthy.

Result 3—Poor similarity; test vector is not a basic factor; predicted values for free-floated points are not trustworthy.

Result 4—Fair similarity; test vector is probably a borderline basic factor, although the predicted values for designees 4 and 6 are suspiciously outside the range of the test points; predicted values for free-floated points are probably not trustworthy.

To illustrate the qualitative approach, several hypothetical results are presented in Tables 6.4 and 6.5. The examples in Table 6.4 involve complete test vectors; the examples in Table 6.5 involve test vectors with free-floated points.

Five hypothetical results from target testing a complete test vector are given in Table 6.4. The error in the points in the test vector is taken to be less than five. Conclusions based on a qualitative comparison of test and predicted vectors for this error are stated at the bottom of the table. Results of the kind illustrated in the third result are common in target factor analysis and should not be rejected routinely. Chemical insight into the nature of the test vector coupled with the conclusions from the quantitative approach discussed above should play the final role in evaluating such borderline results. Although result 4 indicates an unsuccessful test, such results may suggest interesting new test vectors. In this case the test value for designee 2, which is predicted poorly, might be free-floated to form a new target vector.

Incomplete test vectors are the norm in chemical problems since, because of insufficient data or insight, several points usually have to be left blank in a test vector. Four hypothetical results from a target test involving free floating are shown in Table 6.5. Since three points are specified on the test vector and the number of factors is two, the greater-than-n rule is obeyed and the tests are valid.

Qualitative conclusions based on the similarity between the test and predicted vectors are shown at the bottom of the table.

Interpretations of target tests involving free-floated points require special caution, since a test vector becomes more poorly defined as the number of free-floated points increases. Results for the first three tests in Table 6.5 are interpreted using principles invoked for Table 6.4, with the proviso that chemists will have somewhat less confidence in conclusions from tests employing free-floating points. Predicted vector 4 contains two suspicious values: a relatively large value for designee 6 and a negative value for designee 4. Whenever predicted values are outside the range of input test points, the chemist should be wary. Such situations might arise if test vectors are poorly anchored on either the high or low extremes of the complete basic factor. Unless atypical predicted values can be rationalized chemically, the physical significance of the factor should be questioned even though the transformation appears to be successful.

6.8   COMBINATIONS OF BASIC VECTORS

The target-combination procedure can be applied either to selected basic vectors or to all of the successfully transformed basic vectors. Computational facilities and the objectives of the problem will dictate the scope of the combination step using basic vectors.

If all the test vectors that target-transformed with borderline success or better are included in the combination tests, the total number of combinations might be staggeringly large. For example, if 35 vectors passed the target test, there would be, in a six-factor problem, $35 \times 34 \times 33 \times 32 \times 31 \times 30/6! = 1,623,160$ combinations, requiring several hours of execution time even on major computers. Fortunately, many pairs of basic vectors may be equivalent physically, so that all possible combinations do not have to be tested. Scientific criteria can be used to reduce to a manageable size the number of vectors tested in combination. Vectors having the lowest SPOIL values (see Section 4.6) are particularly good candidates for inclusion in combinations. On the other hand, we might eliminate some or all of the vectors which transformed less successfully as well as those which seem equivalent using chemical or TFA criteria.

Complete vectors should be used in combination. Should vectors with free-floated points be employed in combination, the reproduced matrix will lack elements corresponding to each free-floated designee. To circumvent this problem, the corresponding value from the predicted vector is inserted for each missing point. Combination will then produce a complete matrix. However, as more predicted values are employed in combination, the combination solution becomes less reliable. If, inadvertently, only predicted vectors, rather than test vectors, were used in combination, the solution would be an abstract one lacking real input. Such solutions should be discarded.

The chemist should attempt to justify the key combination solutions in terms of theoretical and empirical knowledge. Carefully planned factor analyses based on sound conceptual models pay handsome dividends at this stage of the analysis. Correlating combination-TFA models with chemical insight is the most satisfying end product of target factor analysis. Considering the complexity of most chemical problems, key combination sets of basic vectors do not necessarily have to reproduce the data near experimental error. In a difficult problem a TFA model with a RMS error of, say, three times the experimental error may be considered exceptionally good.

The combination results can be used to identify equivalent basic vectors and to find which basic vectors are represented most often in the better TFA models. Such information is useful for comparing theoretical and empirical models, and for deciding on practical models for data. Methodologies for extracting these kinds of information were explained in Section 6.5.

6.9 PREDICTIONS FROM KEY SETS OF BASIC VECTORS

The procedure for predicting new points in an expanded data matrix using key sets of basic vectors is analogous to the procedure involving typical vectors described in Section 6.6. This last step of a complete target factor analysis can be carried out only if good sets of basic vectors are identified in target combination.

We start with a set of n basic vectors which in combination reproduce the data within reasonable specifications. To predict the data associated with new row designees labeled $a, b, \ldots, m$, values for each of the key, basic cofactors for each new designee are required. For example, suppose that the key combination set in a two-factor problem is shown to involve the two key parameters labeled x and y. We can then use the procedure described in Section 6.6. The column matrix here will have the form

	column designee				
key factor	1	2	3	4	5
x (associated with key parameter x)	c_{x1}	c_{x2}	c_{x3}	c_{x4}	c_{x5}
y (associated with key parameter y)	c_{y1}	c_{y2}	c_{y3}	c_{y4}	c_{y5}

(6.4)

where the cofactors c_{jk} are calculated from target combination involving key parameters x and y. The new row matrix involving the new row designees is

$$
\begin{array}{cc}
\begin{array}{c}
\text{new row} \\
\text{designee} \\
a \\
b \\
\vdots \\
m
\end{array}
&
\begin{array}{c}
\text{key factor} \\
\begin{array}{cc}
2 & 3
\end{array} \\
\begin{bmatrix}
d_{ax} & d_{ay} \\
d_{bx} & d_{by} \\
\vdots & \vdots \\
d_{mx} & d_{my}
\end{bmatrix}
\end{array}
\end{array}
\qquad (6.5)
$$

where the cofactors d_{ij} are known values of the key parameters for the new row designees. Premultiplying column matrix (6.4) by row matrix (6.5) produces a matrix of TFA predicted data. For example, new datum d_{c5} is given by

$$
d_{cx}c_{x5} + d_{cy}c_{y5} = d_{c5}(\text{predicted}) \qquad (6.6)
$$

Data predicted from a combination of basic vectors are expected to have an RMS error at best equal to and probably greater than the RMS error obtained when the data matrix is reproduced using that key set of basic vectors. If the key factors do not span the factor space of the new designees, the RMS error for predicted data will often be quite large. Should a new designee be chemically similar to some of the original row designees, predictions for that designee are more likely to be reliable. Target predictions should be checked whenever possible by comparing predicted values to measured values.

6.10   COMPUTER PROGRAM

A comprehensive computer program for target factor analysis is available from the Quantum Chemistry Program Exchange (QCPE).[8] The initial version of the TFA program was developed by Malinowski and coworkers;[13] the QCPE version called FACTANAL resulted from extensions of Malinowski's program by Howery and coworkers.

The TFA–QCPE program includes detailed input instructions, many options for treating the data and for tabulating results, and documentation for a sample problem. Users of the program need to learn how to input the data matrices and the test vectors, and how to instruct the computer to perform the various stages of target FA. Five kinds of input cards or card decks—job control language (JCL) cards, program deck, control cards, data-matrix deck, and test-vector deck—are employed. Job control language cards depend on the local computer system. These cards initiate and terminate jobs and access FACTANAL from the computer center library. The program deck contains about 2500 cards. Written in FORTRAN IV and checked by IBM Model 370 computers, the program requires a core size of 194K for a 40×40 matrix. Control cards contain instructions for carrying out the desired calculations in a given TFA run (see below). The data matrix and the test vectors are entered according to instructions in the introductory comments for the program.

Test vectors related either to the row designees or to the column designees can be target-tested with the TFA program. Vectors characterizing the row designees are tested directly; vectors associated with the column designees are tested after transposing the data matrix. A complete listing for the input of two runs involving a 6×7 matrix obeying a three-factor model is included in FACTANAL.

Two control cards are the operational heart of the program. A series of options for carrying out all aspects of a complete target factor analysis are incorporated into these two cards. Execution instructions are entered as number or letters in specified spaces on the control cards. Summaries of results from the reproduction step, from uniqueness tests, and from the combination step facilitate the interpretation of analyses. The most up-to-date versions of the program called TARGETFA[11] and BIGFA[14] incorporate direct target testing for column designees and the theory of errors discussed in Chapter 4, and also include summaries of results from target testing and from the prediction step.

We have now completed our general discussion of factor analysis. Applications of factor analysis to chemistry, covering nearly a hundred publications, will be summarized in Chapters 7 through 10.

REFERENCES

1. B. R. Kowalski (Ed.), *Chemometrics: Theory and Application,* ACS Symp. Ser. 52, American Chemical Society, Washington, D.C., 1977.
2. R. F. Hirsch (Ed.), *Statistics,* Franklin Institute Press, Philadelphia, 1978.
3. N. R. Draper and H. Smith, *Applied Regression Analysis,* Wiley, New York, 1966.
4. P. C. Jurs and T. L. Isenhour, *Chemical Applications of Pattern Recognition,* Wiley-Interscience, New York, 1975.
5. A. J. Barr, J. H. Goodnight, J. P. Sall, and J. T. Hewig, *Statistical Analysis System,* Statistical Analysis System Institute, Raleigh, N.C., 1976.
6. N. H. Nie, C. H. Hull, J. G. Jenkins, K. Steinbrenner, and D. H. Bent, *Statistical Package for the Social Sciences,* 2nd ed., McGraw-Hill, New York, 1975.
7. R. J. Rummel, *Applied Factor Analysis,* Northwestern University Press, Evanston, Ill., 1970.
8. E. R. Malinowski, D. G. Howery, P. H. Weiner, J. H. Soroka, P. T. Funke, R. S. Selzer, and A. Levinstone, "FACTANAL," Program 320, Quantum Chemistry Program Exchange, Indiana University, Bloomington, Ind., 1976.
9. E. R. Malinowski, *Anal. Chim. Acta,* **103,** 339 (1978).
10. R. B. Cattell, *The Scientific Use of Factor Analysis in Behavioral and Life Sciences,* Plenum Press, New York, 1978, p. 269.
11. D. G. Howery, G. Williams, and S. Cento, Brooklyn College, Brooklyn, N.Y., unpublished results.
12. R. W. Rozett and E. M. Petersen, *Anal. Chem.,* **47,** 2377 (1975).
13. P. H. Weiner, E. R. Malinowski, and A. Levinstone, *J. Phys. Chem.,* **74,** 4537 (1970).
14. E. R. Malinowski and H. Rozyn, Stevens Institute of Technology, Hoboken, N.J., unpublished results.

7 Component Analysis

In this chapter we describe how investigators have made use of the factor analytical approach for both qualitative and quantitative analysis. The discussion will involve a variety of analytical methods such as absorption and emission spectroscopy, optical rotation, gas chromatography, and mass spectrometry. Our intent is to illustrate the various FA approaches that can be employed for component analysis.

7.1   ABSORPTION SPECTRA

7.1.1   Rationale

Factor analysis has been used in a powerful fashion for determining the number of components that contribute to the absorption spectra of multicomponent systems. This is to be expected because Beer's law, for a multicomponent system, involves a linear sum of product functions:

$$A = \sum_{j=1}^{n_c} \epsilon_j c_j \tag{7.1}$$

Here A is the absorbance at a given wavelength measured in a cell of unit path length, ϵ_j the extinction coefficient of the jth component, and c_j the concentration of component j, the sum being taken over all n_c components in the mixture. Factor analysis is ideally suited to systems where the absorbance spectra of each individual component differ significantly. The fact that the extinction coefficients need not be unique at all wavelengths is one of the advantages of factor analysis over other methods which involve simultaneous equations.

For a series of solutions having the same species but different concentrations, the absorbances are best expressed as

$$A_{\lambda k} = \sum_{j=1}^{n_c} \epsilon_{\lambda j} c_{jk} \tag{7.2}$$

Here λ refers to the wavelength, k refers to the particular solution, and j refers to the component. These equations can be written in a more compact form utilizing matrix notation:

$$[A] = [E][C] \tag{7.3}$$

If there are n_c absorbing components in n_s solutions and the measurements are made at n_w wavelengths, then $[A]$ is an $n_w \times n_s$ absorbance matrix, $[E]$ is an $n_w \times n_c$ extinction coefficient matrix, and $[C]$ is an $n_c \times n_s$ concentration matrix.

The absorbance data matrix is obtained by digitizing the spectra at intervals over a wide wavelength range. In this way, the entire band contour is analyzed rather than only the major spectral features. The use of a large number of digitized wavelengths is desirable, since this will yield statistically more accurate results. The number of solutions of different composition to be used in the factor analysis study usually poses a more time-consuming problem. However, the larger the number of solutions studied, the more reliable the results. To ensure that the rank of the data matrix, as determined by factor analysis, will equal the number of components, both the number of digitized wavelengths and the number of solutions must not be less than the number of components. It is relatively easy to use several hundred digitized wavelengths, whereas 10 solutions are usually considered sufficient when there are about six or fewer components.

Because of experimental error, factor analysis will always yield a set of eigenvectors whose number will be equal to the number of columns in the data matrix. However, as explained in Chapter 4, eigenvectors associated with small eigenvalues simply reproduce experimental error and should be disregarded. A variety of statistical tests have been developed specifically to estimate the rank of a data matrix. These methods, discussed in Chapter 4, include (1) comparing the misfit between the experimental absorbance matrix and the absorbance matrix regenerated using different numbers of abstract factors; (2) comparing the residual standard deviation, calculated from the eigenvalues, with the estimated standard deviation; (3) disregarding those eigenvectors associated with eigenvalues which are smaller than their standard errors; (4) using the χ^2 criterion; (5) examining the imbedded error function; and (6) studying the factor indicator function. It is good practice to use a combination and these methods to deduce the dimensionality of the factor space, which, in these problems, is equivalent to the number of components in the mixtures.

7.1.2 Pioneering Efforts

Although the early investigators did not use factor analysis to determine the number of components, their pioneering efforts laid the foundation for its ultimate utilization. For this reason we will briefly review the early methodology in chronological order.

In 1960, Wallace[1] was the first to recognize that rank analysis of absorption spectra could be used to determine the number of components in a mixture (the first step of factor analysis). He recorded, at 25-nm intervals, the visible spectra of methyl orange and methyl red indicators, as well as mixtures of the indicators, in buffered solutions of known pH. To determine the rank of the A matrix, Wallace did not employ the sophisticated factor analysis procedure described in Chapter 3. Instead, he used a simple statistical criterion which states that the determinant of a singular matrix is equal to its standard deviation. A square singular matrix is one in which the number of rows, or columns, equals the rank plus one. The rank of the data matrix equals the order of the largest nonsingular submatrix. The problem then is to find this submatrix.

The rank of each A matrix was deduced in the following way. First every possible 2 × 2 submatrix that could be formed from the A matrix was constructed. The value of the determinant, d, for each 2 × 2 submatrix was then compared to the standard deviation, σ. If the determinant was close to the standard deviation, the submatrix was deemed singular. If one of the 2 × 2 submatrices was nonsingular, as proven by $d > \sigma$, every possible 3×3 submatrix was constructed by adding a row and a column to this submatrix. This procedure of adding rows and columns was repeated until the largest singular matrix was found. The rank was taken to be the number of rows (or columns) in this matrix. Using this technique, Wallace concluded that solutions of supposedly pure methyl red and pure methyl orange each contain two absorbing components. This technique was also applied to data concerning mixtures of the two dyes. Rank analysis showed that there were four components present in the mixtures.

Although this method is readily applicable to small data matrices, it is unwieldy when large matrices are involved. Furthermore, the method leads to a dead end in the overall FA scheme because it can yield only one piece of information, the rank of the data matrix, and nothing more. However, this pioneering investigation brought to the attention of the chemist the fact that matrix rank analysis is a viable method for determining the number of absorbing components in a series of related mixtures.

Because this method is excessively tedious to carry out when large data matrices are involved, Wallace and Katz[2] developed the following alternative method. In addition to the absorbance matrix, A, an error matrix, S, was constructed. Each element of the error matrix is simply the estimated error associated with the corresponding element of the absorbance matrix. Using standard mathematical procedures[3] the A matrix is reduced to a matrix whose elements are all zero below the principal diagonal. At each step in the reduction of A, appropriate operations, based on statistical theory, are performed on the error matrix, yielding a reduced S matrix. The rank of the A matrix is equivalent to the number of diagonal elements in the reduced A matrix, with absolute values greater than three times the corresponding elements in the reduced S matrix.

The original A matrix consisted of the absorbances of a methyl red solution as a function of wavelength and pH. Because the concentration of an absorbing ligand is a function of pH, varying the pH caused a change in absorbance. The original S matrix consisted of 64 identical elements, all equal to the experimental error, 0.003. Comparing the reduced A matrix to the reduced S matrix, Wallace and Katz[2] concluded that there were at least three absorbing components present, and possibly four, since one element in the reduced A matrix was approximately three times the corresponding element in the reduced S matrix.

Varga and Veatch,[4] relying upon the computer method of Wallace and Katz,[2] investigated the nature and stabilities of hafnium chloranilic acid metallochrome. Two series of solutions were prepared, one containing chloranilic acid at various concentrations and the other containing both hafnium(IV) and chloranilic acid. The concentrations of chloranilic acid in the first set of 12 solutions were identical to the second set, which contained both hafnium and chloranilic acid. In the second set of 12 solutions, the total molar concentration of hafnium plus chloranilic acid was held constant. Absorbance measurements were made at 5-nm intervals from 260 to 360 nm over the region of maximum absorption. Thus two 21 × 12 individual data matrices were investigated. Their ranks were determined by comparing the reduced absorbance matrices with their reduced error matrices. They concluded that three absorbing species were present in the hafnium–chloranilic acid solutions.

In deducing the rank, a knowledge of the size of the experimental error is extremely important. Varga and Veatch[4] recognized this fact and made a systematic study of the effect of error on both the chloranilic acid and hafnium–chloranilic acid systems. When the overall absorbance error estimate was varied from 0.003 to 0.050, for chloranilic acid, the rank changed from 5 to 1. Since only one species, undissociated chloranilic acid, is expected to exist in 3 M perchloric acid solution, the rank should be 1. This is consistent with the results if the error is estimated to be about 0.025 absorbance unit. The rank increases dramatically when the error is assumed to be less than 0.010 absorbance unit, reflecting the sensitivity to random fluctuations in absorbance measurements rather than to the number of absorbing species present.

A similar situation exists for the hafnium–chloranilic acid mixtures. The accepted rank is three if the error is estimated to be 0.025 unit. If the error is assumed to be less than 0.016 unit, the rank fallaciously increases. If it is assumed to be greater than 0.040, the rank fallaciously decreases.

Because of this sensitivity, Varga and Veatch[4] recognized that the assumption of a constant absorbance error could lead to incorrect conclusions. They calculated the absorbance error, $S_{\lambda k}$, from the photometric error in the transmittance, ΔT, and the measured absorbance, $A_{\lambda k}$:

$$S_{\lambda k} = 0.43429 \cdot \Delta T \cdot \text{antilog } A_{\lambda k} \qquad (7.4)$$

Using this equation together with the experimental absorbances for the hafnium–chloranilic acid solutions, they generated a series of error matrices based upon different estimates of ΔT ranging from 0.001 to 0.050. Again the conclusions concerning the ranks of the matrices were found to depend upon an accurate estimation of the photometric error. For the Beckman DU spectrophotometer employed, the photometric error was in the range between 0.3% and 0.5% ($\Delta T = 0.003$ to 0.005). This led to the conclusion that there were three species present.

Katakis[5] developed a computer method for determining the rank of an absorbance matrix, based upon the Gauss process of elimination.[6] Here the A matrix is reduced by subtracting an appropriate matrix constructed from the elements of the row and column associated with the largest element in the matrix. This subtraction is repeated until all the elements of the residual matrix are less, in absolute value, than the corresponding elements of the error matrix. The rank of A equals n, the number of matrix subtractions required. Katakis applied this method to study the absorbance of Cr^{2+} solutions.

A graphical method for determining the rank of the absorbance matrix was developed by Coleman et al.[7] This method relies upon the same numerical relationships used previously but does not require sophisticated computer analysis. The necessary computations can be done rapidly with a desk calculator. When applied to the chloranilic acid and methyl red spectra previously discussed, the same conclusions were reached.

All the methods just described represent variations in different rank analysis techniques. These studies are included here to stress the historical sequence and the importance of rank analysis in spectrophotometric studies of multicomponent systems. Rank analysis is the first step in factor analysis.

7.1.3 Factor Analyses

Kankare[8] recognized that factor analysis was an ideal mathematical tool for determining the number of components in a solution from its absorption spectrum. He was inspired by Simmonds,[9,10] who applied factor analysis to optical response in photography; by Reeves,[11] who used factor analysis to separate medium effects from concentration effects in dye solutions; and by Wernimont,[12] who used factor analysis to evaluate the performance of different spectrophotometers (see Section 10.3).

He recorded the absorption spectra of solutions containing 8×10^{-5} M bismuth ion, 1 M perchloric acid, and varying amounts of sodium chloride and sodium perchlorate to maintain a constant ionic strength. The spectra were recorded against blanks having the same composition but no bismuth. Absorbance measurements were made on 17 solutions between 230 and 360 nm at 5-nm intervals.

To determine the rank of the absorbance matrix, Kankare compared the residual standard deviation with the estimated deviation. This method is described in Chapter 4. Since the residual standard deviation calculated from factor analysis must be less than the estimated deviation, he concluded that seven absorbing species were present. These species were suspected to be Bi^{3+}, $BiCl^{2+}$, $BiCl_2^+$, $BiCl_3$, $BiCl_4^-$, $BiCl_5^{2-}$, and $BiCl_6^{3-}$.

Kankare[8] used FA to improve the data by substituting factor analysis-regenerated data points for all points with excessive errors. A data point was considered to have an excessive error if the absolute value of the difference between the measured absorbance and the AFA regenerated absorbance was greater than three times the standard deviation. Such errors were considered to be accidentally excessive. Thus data points with large errors were easily spotted and removed. The smoothed data matrix was then factor-analyzed, yielding, hopefully, more reliable results.

Factor analysis of the absorbance matrix does not yield directly an extinction coefficient matrix $[E]$ and a concentration matrix $[C]$ in a true chemical sense, as portrayed by (7.3). Instead, it yields mathematical solutions which have basis axes (eigenvectors) which lie in the same chemical space, but do not necessarily coincide with the chemical axes. These mathematical axes must be transformed into chemical axes. Additional information is required to perform such transformations.

Kankare[8] was able to obtain the concentration matrix by first speculating what the seven absorbing bismuth species were and then using, as a first approximation, the formation constants for these species as determined by other methods. He then developed a least-squares method for the purpose of adjusting these constants to give the best set compatible with the abstract matrices of factor analysis. The formation constants from the factor analysis study were considered to be more accurate and reliable.

Knowing $[A]$ and having obtained $[C]$, Kankare readily calculated $[E]$. The elements of each column of $[E]$ trace out the spectrum of one of the absorbing bismuth species, even though it is impossible to obtain the spectra of these species by direct spectrophotometric measurement. With the aid of factor analysis, the spectrum of each absorbing species in the extremely complicated mixture was obtained.

Factor analysis of absorption spectra was used by Hugus and El-Awady[13] in their investigation of the hydrolytic depolymerization of certain binuclear cobalt(III) complexes. To test their detailed kinetic model, they needed to know how many absorbing species were present. Their data matrix consisted of the absorbances of 38 solutions measured at nine wavelengths. Four criteria for deducing the number of species were employed: (1) trends in the eigenvalue, (2) the standard error in the eigenvalue, (3) the number of residuals greater than three times the estimated standard deviation, and (4) the chi-squared test. These criteria are fully discussed in Section 4.3. The results shown in Table 7.1 indicate that three species are present. As shown in the table, there is a severe drop in the eigenvalue when n is changed from 3 to 4, indicating three factors. The first three eigenvalues are greater than their respective standard errors, whereas the remaining six eigenvalues are much smaller than their standard errors. Of the 342 data points, 155 reproduced data points have an error greater than three times the estimated error when two factors are employed. When three factors are employed, all reproduced data points have errors less than three times the estimated error. The expectation value $[\chi_n^2(\text{expected}) = (r - n)(c - n)]$ using two factors is much smaller than the calculated χ_n^2, whereas the expectation value using three factors is closer to χ_n^2. All these criteria give evidence that three species are present.

By studying the imbedded error function (IE) and the factor indicator function (IND) (see Section 4.3), Malinowski[14,15] substantiated the conclusion that three components are present. As shown in Table 7.1, there is no further reduction in IE upon using more than three eigenvectors. Furthermore, the IND function

Table 7.1   Factor analysis study of the absorbances of a Co(III) complex and its hydrolysis products[a]

n	λ_n (Eigenvalue)	Standard Error in λ_n	3σ misfit	X_n^2 Calculated	X_n^2 Expected	Imbedded Error IE	Indicator Function, IND × 10⁵	Real Error, RE
1	1,627.301311	0.231	311	222,227	296	0.03219	150.90	0.09658
2	2.642417	0.092	155	8,742	252	0.01083	46.87	0.02297
3	0.140080	0.111	0	35	210	0.00060	2.89	0.00104
4	0.000091	0.057	0	21	170	0.00060	3.62	0.00091
5	0.000057	0.063	0	14	132	0.00060	5.04	0.00081
6	0.000035	0.071	0	10	96	0.00061	8.32	0.00075
7	0.000027	0.056	0	5	62	0.00062	17.45	0.00070
8	0.000022	0.073	0	3	30	0.00063	62.80	0.00063
9	0.000015	0.052	0	0	0	—	—	—

[a] Reprinted with permission from Z. Z. Hugus, Jr. and A. A. El-Awady, *J. Phys. Chem.,* **75,** 2954 (1971), and from E. R. Malinowski, *Anal. Chem.,* **49,** 612 (1977).

reaches a minimum at $n = 3$, again indicating three factors. For $n = 3$, the real error (RE) is 0.00104 absorbance unit. Hence without recourse to any knowledge of the experimental error, as required by the previous criteria, Malinowski deduced not only the number of components but also the experimental error.

The data of Wallace and Katz[2] concerning methyl red solutions as a function of pH, discussed in Section 7.1.2, was factor-analyzed by Hugus and El-Awady.[13] All four error criteria gave evidence that there were only three components. These calculations were based upon the same reasonable error estimation as given by Wallace and Katz. The questionable fourth component mentioned by Wallace and Katz was clearly ruled out by this study. Malinowski,[14] however, presented evidence that the error estimation used to reach this conclusion was too large. He argued that there were four factors involved because the IND function exhibited a minimum at $n = 4$, yielding a real error of 0.00154 absorbance unit, considerably less than 0.003 used by the previous investigators.

Principal factor analysis of the Fourier transform infrared spectra, recorded and digitized from 500 to 3500 cm⁻¹, was used by Rasmussen and coworkers[16] to correctly identify the number of components in artificial mixtures of xylenes. However, the method was unsuccessful for determining the number of components in mixtures of alkanes because the infrared spectra of the individual components were not distinguishable. Malinowski and McCue[17] used target factor analysis of ultraviolet spectra in the region from 260 to 280 nm not only to chemically identify the components in mixtures of xylenes but also to determine the concentrations of the components.

Antoon et al.[18] used factor analysis to investigate the FT–IR spectra of polymeric films. A factor analytical study of the fingerprint region (1100 to 1800 cm^{-1}) of seven films consisting of various proportions of atactic polystyrene and poly-2,6-dimethyl-1,4-phenyl oxide showed that three species were present, giving evidence that one of the constituent polymers undergoes a conformational change which is a function of the compositional blend. An FA–FT–IR study[18] of semicrystalline poly(ethylene terephthalate) yielded two components, corresponding to crystalline and amorphous phases, the trans and gauche conformers being indistinguishable. An FA–FT–IR study[18] of poly(vinyl chloride) films that have been subjected to various annealing treatments gave evidence for the existence of as many as eight components, due to a combination of configurational and conformational disorders in the chains.

Bulmer and Shurvell[19] employed factor analysis as a complement to band resolution studies of infrared spectra. Unfortunately, band resolution techniques require an a priori assumption concerning the general contour of the band. The shape is usually considered to be Lorentzian. A major advantage of factor analysis is that no such assumption is required.

Bulmer and Shurvell recorded and factor-analyzed the infrared spectra of the carbonyl region of acetic acid–CCl$_4$ solutions. Their purpose was to study the monomer–cyclic dimer equilibrium. Their data matrix was constructed as follows. The spectra were digitized manually every 0.5 cm^{-1} from 1690.5 to 1790.0 cm^{-1}. The absorbance matrix consisted of 200 data points from each of nine solutions, ranging from 1.72×10^{-4} to 4.31×10^{-2} M acetic acid in CCl$_4$ solvent. The standard error in the eigenvalue, the $3\ \sigma$ misfit, and the chi-squared criteria clearly indicated that not two but four absorbing components were present. This conclusion was substantiated by Malinowski,[14,15] who studied the IE and IND functions. The results of these studies are discussed in Section 4.3 (see Table 4.4). The confirmation of the existence of four species invalidated all previous approaches, which assumed only two species, a monomer and cyclic dimer of acetic acid. The factor analytical studies provided evidence for the existence of other hydrogen-bonded species.

Because of the success of the factor analysis technique in establishing the number of components in acetic acid solution, Bulmer and Shurvell[20] investigated trichloroacetic acid. They recorded the infrared spectra of nine solutions of trichloroacetic acid in CCl$_4$ solvent ranging from 0.61×10^{-3} to $0.16\ M$. At all concentrations, only two bands visually appeared in the spectrum of the carbonyl region. However, factor analysis gave evidence that, as in the case of acetic acid, four components were actually present. The detection of species other than monomer or dimer in this system cannot be gotten from simple band contour analysis nor from monomer–dimer equilibrium constants calculated from concentration studies.

The factor analysis technique was applied by Bulmer and Shurvell[21] to investigate the infrared spectra of solutions of $CDCl_3$ and di-*n*-butyl ether in CCl_4. Infrared spectra were recorded from 2290 to 2310 cm^{-1}, a region characteristic of the C–D stretching of $CDCl_3$. Nine solutions were prepared, in which the concentration of $CDCl_3$ was held constant throughout at approximately 2.2 *M* and the concentration of di-*n*-butyl ether was varied from about 0.15 to 1.00 *M*. The spectra were digitized into 241 wavenumbers. Thus a 241 × 9 data matrix, containing 2169 points, was obtained.

The results of the eigenvalue study from factor analysis clearly showed that only two components were present. If this were so, then, in accord with the traditional approach, an isosbestic point should have occurred when the spectra were normalized to unit concentration and unit path length. No such intersection was obtained. Because factor analysis indicated only two components, Bulmer and Shurvell[21] searched for an explanation for the dilemma. Further investigation revealed that an isosbestic point occurred when the spectra were normalized to unit area as well as unit concentration and unit path length.

To obtain the extinction coefficient and concentration matrices, [*E*] and [*C*], Bulmer and Shurvell[19] applied the method developed by Kankare[8] described earlier. First, a rough estimate of the equilibrium constant for the formation of the 1:1 complex between chloroform and di-*n*-butyl ether was made. The equilibrium constant was then varied until the abstract concentration matrix from FA was compatible with the corresponding matrix obtained using the equilibrium constant. The constant 0.0431 M^{-1} obtained in this way agreed very well with the value 0.0456 M^{-1} obtained from band resolution studies.

Korppi-Tommola and Shurvell[22] studied the complex formation between pentachlorophenol (PCP) and acetone in CCl_4 solution by factor analyzing, separately, the carbonyl and the hydroxyl stretching regions in the infrared. The stoichiometric concentration of PCP was fixed at 0.05 *M* for one study and at 0.10 *M* for another study, while the acetone concentration was varied from 0.005 to 1.0 *M*. The absorbance matrix for the carbonyl region consisted of 100 digitized wavelength intervals for each of eight different solutions. For the hydroxyl region 401 digitized wavelength intervals per spectrum were involved.

Based upon various criteria, such as chi-squared and residual standard deviation in the eigenvalues, factor analysis indicated three components for both types of data matrices. However, each region was interpreted quite differently. The hydroxyl region was believed to be the result of one monomer and two complexes, whereas the carbonyl region was believed to be due to one monomer, one combination band, and only one complex. Such an anomoly could occur if either one of the complexes did not significantly affect the carbonyl vibration or if both complexes contributed identically to the carbonyl vibration.

Although band resolution studies, based upon Cauchy–Gauss product functions, for the hydroxyl region yielded a better spectral fit when four components were used, factor analysis gave clear evidence for only three absorbing components. Thus factor analysis served as an excellent complement to band resolution studies because it reduced the temptation to add an excessive number of bands in order to improve the fit.

Band resolution studies require some assumption of the band shape, such as Gaussian or Lorentzian. In actuality, the assumed shape may be incorrect. Lawton and Sylvestre[23] proposed a method, based upon factor analysis, which requires no such assumption and which permits overlapped bands to be separated into their true shapes. This *self-modeling method* requires regions of the spectra where each component individually absorbs radiation.

Their study focused attention on a set of mixtures of standard dyes resulting from a production process. The absorption spectra of five mixtures, observed in the visible region, are shown in Figure 7.1. Principal factor analysis was formed on an absorbance matrix constructed by digitizing the wavelength scale

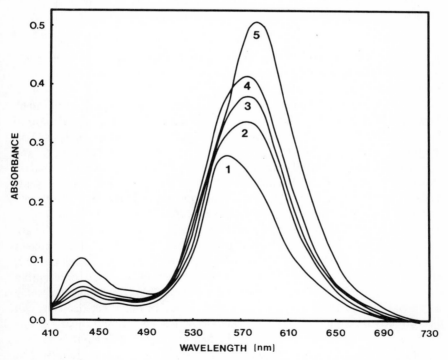

Fig. 7.1. Visible spectra of five mixtures of standard dyes [Reprinted with permission from W. H. Lawton and E. A. Sylvestre, *Technometrics,* **13,** 617 (1971)].

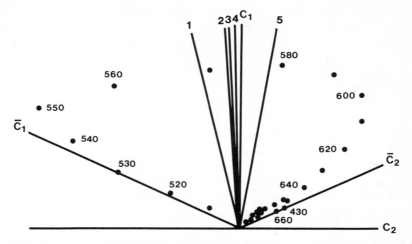

Fig. 7.2. Results of principal factor analysis of spectra shown in Figure 7.1, after digitization (based upon work of W. H. Lawton and E. A. Sylvestre, *Technometrics*, **13**, 617 (1971)].

every 10 nm from 410 to 700 nm. Since the first two principal eigenvectors accounted for the data within experimental error, it was concluded that the mixtures were composed of only two dyes.

A mathematical method for isolating the spectra of each dye was devised by Lawton and Sylvestre.[23] The rationale of the method can be easily visualized by examining Figure 7.2, which shows the results of the factor analysis. This diagram, analogous to Figure 5.2, shows the normalized principal axes, C_1 and C_2, and the five normalized data axes, labeled 1, 2, 3, 4, and 5, where each data axis represents one of the five mixtures. Each point on the diagram represents a specific wavelength. The chronological, oblique projection of these points onto any data axis virtually traces out the visible spectra of that mixture. According to Beer's law, the absorbance of a mixture is a linear combination of the absorbance of its pure components. Because the absorbance of a pure component is a product of its absorptivity and concentration, both quantities being positive, all the data points must lie in a region between a set of real axes representing the pure components. If the individual components absorb radiation of some wavelength uniquely, the pure component axes can be found because they will pass directly through the unique wavelength points. As seen in Figure 7.2, one such axis, $\overline{C}_1$, can be drawn through the 530-nm wavelength point, while the second, $\overline{C}_2$, can be drawn through the 430- and 660-nm points jointly. The spectrum of each pure component can now be generated by projecting, obliquely, the wavelength points onto the appropriate component axes. When this was done,

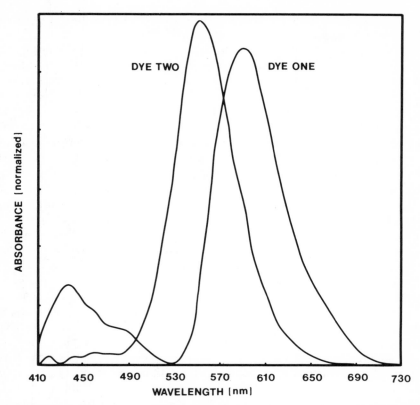

Fig. 7.3. Spectra of pure components generated by projecting the wavelength points shown in Figure 7.2 onto the respective pure component axes [Reprinted with permission from W. H. Lawton and E. A. Sylvestre, *Technometrics,* **13,** 617 (1971)].

the spectra shown in Figure 7.3 were resolved. Unfortunately, from mixed spectra alone it is impossible to determine the relative composition of the components.

Although the self-modeling method was originally restricted to two-component systems, Ohta[24] expanded the technique to a three-component system involving cyan, magenta, and yellow dyes, which are used in subtractive color photography. Although the spectral density distribution curves of the three dyes could not be determined uniquely, their ranges could be restricted to certain limits because the spectral densities and the spectral buildup of the mixtures is always nonnegative. Sylvestre et al.[25] showed that, under special circumstances involving chemical equilibria and reaction kinetics, the self-modeling technique could be applied to multicomponent systems beyond two or three components, yielding resolved spectra for each of the components.

When a polyelectrolyte is added to an aqueous solution of a staining dye, a new absorption band often appears at a lower wavelength than that of the free dye. A factor analytical study of this phenomenon, called *metachromasy,* was conducted by Yamaoko and Takatsuki.[26] The visible spectra of two metachromatic dyes, crystal violet and trypaflavine, in the presence of seven different polyelectrolytes were measured at various concentration ratios of dye and polyanion. Factor analysis of the 14 data matrices revealed only two absorbing components in each case. The two components were interpreted to be the free dye and the dye molecule bound to the polyanion. The spectra of the bound dyes were generated by the spectral isolation method of Lawton and Sylvestre.[23]

Lin and coworkers[27,28] developed a factor analytical method, called *automated spectral isolation* (ASI), which they successfully used to isolate component spectra from the spectra of mixtures. The unique feature of ASI is that it does not require regions of spectral purity. The technique consists of the following. First each digitized spectrum is normalized so that the absorbance of the maximum is 1.000. PFA is then used to define the primary eigenvectors and, hence, the number of components. To find the spectral axes of the pure components, "prototype spectra" are target-tested and the predicted vectors are judged by means of a *risk function.* The smaller the risk function, the closer the regenerated prototype spectrum is expected to resemble a pure component spectrum.

Prototype spectra for infrared are generated by a single-needle search which is similar to the uniqueness test but makes use of finite absorbance values of 0.08 instead of zeros. In other words, the test vectors consist of (1.000, 0.08, 0.08, ..., 0.08), (0.08, 1.000, 0.08, ..., 0.08), ..., (0.08, 0.08, 0.08, ..., 1.000). The purpose of using a base of 0.08 instead of 0.0 is to give better assurance that the absorbance values of the predicted target spectra will be positive rather than negative, which is physically unrealizable. The predicted target spectra that result from each of these prototype test vectors are considered as possible candidates for the pure component spectra. The risk value of each prototype spectrum is computed by the following defining equation:

$$(RISK)_j = \frac{1}{r} \sum_{i=1}^{r} (s_{ij} - \bar{\bar{r}}_{ij})^2 \sum_{i=1}^{r} s_{ij}^2 \qquad (7.5)$$

Here $\bar{\bar{r}}_{ij}$ is the ith absorbance value in the jth test vector and s_{ij} is the corrected ith absorbance $\bar{r}_{ij}$ in the predicted spectrum,

$$s_{ij} = \begin{cases} \bar{r}_{ij} \cdot & \text{if } \bar{r}_{ij} \geq 0 \\ 10\bar{r}_{ij} & \text{if } \bar{r}_{ij} < 0 \end{cases} \qquad (7.6)$$

The sum is taken over all r components of the test vector. The reason for using

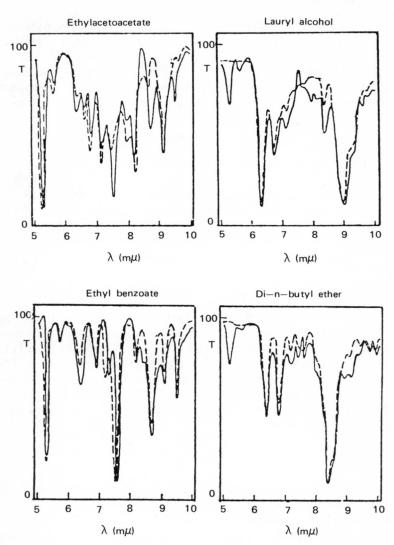

Fig. 7.4. Isolated spectra (solid lines) and true component spectra (dashed lines) [Reprinted with permission from C. M. Lin and S. C. Liu, *J. Chin. Chem. Soc.,* **25,** 167 (1978)].

s_{ij} instead of $\bar{r}_{ij}$ is to intuitively increase the risk value when negative absorbances appear in the target-reproduced prototype spectrum. Although there are r prototype spectra, only n of them, having the smallest risk values, will yield predicted spectra corresponding to the pure components.

Lin and coworkers[27] recorded the infrared spectra of ethylacetoacetate, lauryl alcohol, ethyl benzoate, and di-*n*-butyl ether. Using Beer's law, they generated artificial spectra of eight mixtures containing different amounts of these components. The eight-mixture data matrix was subjected to the ASI technique. Figure 7.4 shows the spectrum of each pure component and the isolated spectra obtained from the set of nonredundant prototype spectra having the smallest risk values.

7.2 EMISSION SPECTRA

Factor analysis has proven to be a unique aid for determining the number of components responsible for an emission spectrum. In methods such as Raman and fluorescence spectroscopy, the emission intensities depend upon the concentrations as well as the unique spectral properties of each emitting species. Consequently, factor analysis is readily applicable.

7.2.1 Raman

Factor analysis was successfully applied to the laser Raman spectra of aqueous indium(III) chloride solutions by Jarv et al.[29] A single, broad, asymmetric Raman band profile was observed, with a maximum that shifted from 311 to 279 cm^{-1} as the chloride-to-indium concentration ratio, R, was increased. This suggested that the single band might be a composite of several indium–chloride ion species. This problem is ideal for factor analysis, since the Raman intensity, I_{ik}, observed at the ith wavenumber for the kth solution obeys the following expression, analogous to Beer's law:

$$I_{ik} = \sum_{j=1}^{n} J_{ij}C_{jk} \qquad (7.7)$$

where J_{ik} is the molar intensity of the jth species at the ith wavenumber and C_{jk} the concentration of the jth species in the kth solution.

For each of 31 solutions of varying R, the Raman spectrum was recorded in digital format. Each spectrum was scanned between 170 and 410 cm^{-1}, corrected for background, and digitized into 481 points. Factor analysis of the resulting 481 × 31 data matrix gave evidence that four species were present. This conclusion was verified by examining the residual standard deviation, chi-squared, the number of misfits greater than four times the standard deviation, and the standard error in the eigenvalue.

Factor analysis was then used to estimate the range of R where less than four components were present. The method gave evidence that only two species were present when R was less than 2.36. This disagreed with other estimations based upon the semi-half-band widths and the incomplete third moments, which indicated three species in this range. The discrepency could be the result of an accidental linearity between the intensities of two of the three species in this region, making these species indistinguishable by factor analysis.

The following four species were postulated to be responsible for the spectra: $[InCl(H_2O)_5]^{2+}$, $[InCl_2(H_2O)_4]^+$, $[InCl_3(H_2O)_3]$, and $[InCl_4(H_2O)_2]^-$. Because of limited accuracy and very severe band overlap, the equilibrium constants between these species could not be calculated. This study shows that the observation of a single band in a Raman spectrum is not sufficient evidence for the existence of a single species.

Raman spectra of aqueous mixtures of $ZnCl_2$ and HCl were studied by Shurvell and Dunham.[30] Spectra were recorded in the Zn–Cl stretching regions using various concentration ratios of chloride and zinc ions. Factor analysis showed that only two light-scattering components existed. These were postulated to be $ZnCl_2$ and $ZnCl_4^{2-}$. Using band resolution techniques, the equilibrium constant for the reaction $ZnCl_2 + 2Cl^- \rightleftharpoons ZnCl_4^{2-}$ was estimated to be 0.22 M^{-2}.

7.2.2   Fluorescence

Nearly all the applications described thus far in this chapter require data matrices involving not one but a series of mixtures composed of varying amounts of the same components. The exceptions discussed earlier require that the components exist in chemical equilibrium so that a change in temperature or pH can produce a change in the composition. A method for determining the number of fluorescent components and their individual fluorescence spectra in a *single* mixture has been aptly demonstrated by Warner and coworkers.[31] Equilibrium between the components is not required for the factor analysis/fluorescence technique. However, the complete fluorescence spectrum of a single component in a mixture can be obtained only when the data matrix includes wavelength regions at which only that component absorbs and emits. The analysis utilizes the fact that each fluorescent component is characterized by a unique dependence of its fluorescent intensities upon two distinct parameters, the excitation wavelength λ_i and the observed emission wavelength λ_j. The data matrix consists of an excitation–emission matrix, $[M]$, whose elements, M_{ij}, are the fluorescent intensities measured at λ_j when the excitation is at λ_i. For dilute mixtures, these intensities depend upon the sum of product functions associated with each fluorescent component, k:

$$M_{ij} = \sum_{k=1}^{n} \alpha_k X_{ik} Y_{kj} \qquad (7.8)$$

where α_k is proportional to the concentration of component k, X_{ik} is proportional to the number of photons absorbed at wavelength λ_i per unit concentration of k, and Y_{kj} is proportional to the fraction of fluorescence emitted by k at wavelength λ_j. Note that X_{ik} is independent of λ_j and Y_{kj} is independent of λ_j. This expression is ideal for factor analysis.

In a series of trial studies, Warner and coworkers[31] factor-analyzed 10 two-component excitation–emission matrices involving five aromatic hydrocarbons: anthracene, pyrene, perylene, chrysene, and fluoranthene. Each data matrix was obtained in the following way. First, they held the excitation wavelength constant while the emission spectrum was scanned. The emitted spectrum was then digitized into 50 wavelengths. The intensities at these wavelengths were formatted and transmitted directly to a computer. The excitation wavelength was then changed and the scanning and digitizing procedures repeated at the same 50 wavelengths. This was repeated 50 times until 2500 data points were acquired. Each scan produced a row of the data matrix. Thus the data formed a 50 × 50 excitation–emission matrix.

A stray light component was found by recording the fluorescence of the pure solvent. The stray light contribution and the estimated dark current were first subtracted from each data matrix. Then a multiple of this stripped matrix was subtracted, to correct for the scattered light contribution. These pretreated data matrices were then subjected to AFA. The rank of each of the 10 different data matrices correctly equaled the number of fluorescent components. The fluorescence spectra of the pure components were deduced without recourse to any a priori knowledge of the pure components or their spectra.

The conventional procedure for determining the quantitative composition of a fluorescent mixture involves fitting the data to a set of simultaneous equations. This procedure requires that we know the identity and the individual fluorescence spectra of all species in the mixture. Ho et al.[32] developed a rank annihilation method which yields the quantitative composition of a single fluorescent species in a mixture without requiring the identification of the other fluoescent components. The basis of the rank annihilation method is the following. The excitation–emission matrix, $[M]$, of a multicomponent mixture has a rank that equals n_c, the number of components present. The rank of the corresponding excitation–emission matrix, $[N]$, of a pure component ideally equals unity. If we subtract the correct amount of $[N]$ from $[M]$, we will obtain a reduced matrix, $[L]$, which would have a rank equal to $n_c - 1$. The amount of $[N]$ that must be subtracted from $[M]$ to accomplish this task is equal to $(c_k/c_k^0)[N]$, where c_k^0 is the concentration of pure component k, in the same solvent, used to obtain $[N]$. In other words,

$$[L] = [M] - \frac{c_k}{c_k^0} [N] \tag{7.9}$$

Even when both data matrices, $[M]$ and $[N]$, are corrected for dark current and light scattering by the solvent, random noise tends to confuse the rank reduction process. An efficient way of determining the correct c_k/c_k^0 value is to

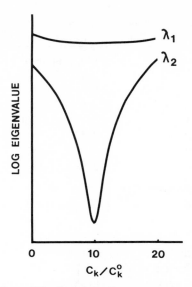

Fig. 7.5. Results of rank annihilation of the fluorescence spectra of a binary mixture of perylene and anthracene [Reprinted with permission from C. N. Ho, G. D. Christian, and E. R. Davidson, *Anal. Chem.*, **50**, 1108 (1978)].

examine the n_cth eigenvalue of $[L]$ as a function of the ratio c_k/c_k^0. The n_cth eigenvalue will reach a minimum at the correct ratio value (see Figure 7.5). This method was successfully applied to mixtures of perylene and anthracene.[32]

7.2.3 Auger

The Fourier-transformed carbon Auger spectra of thin films of polyethylene and five poly(alkyl methacrylates) (alkyl = methyl, ethyl, isobutyl, *n*-butyl, and octadecyl) were recorded and factor-analyzed by Gaarenstroom.[33] Two factors accounted for 99.8% of the variance. Semiquantitative analysis of the polymer films were obtained by treating the polymers as mixtures of polyethylene and poly(methyl methacrylate). The two eigenvectors were target-transformed into axes corresponding to these two components. The factor (composition) loadings obtained from TFA, as shown in Table 7.2, agreed reasonably well with the known mixture fractions.

7.3 KINETICS

Determining the number of reacting species and their concentrations as a function of time is the basis of chemical kinetics. Factor analysis has been ex-

Table 7.2   Composition of polymeric films determined by target factor analysis compared to known fractions[a]

Polymer Film	TFA		Known	
	PE	PMMA	PE	PMMA
Poly(methyl methacrylate) (PMMA)	0.00	1.00	0.00	1.00
Poly(ethyl methacrylate)	0.19	0.82	0.17	0.83
Poly(iso-butyl methacrylate)	0.12	0.88	0.37	0.63
Poly(n-butyl methacrylate)	0.20	0.80	0.37	0.63
Poly(octadecyl methacrylate)	0.65	0.36	0.77	0.23
Polyethylene (PE)	1.00	0.00	1.00	0.00

[a] Reprinted with permission from S. W. Gaarenstroom, *J. Vac. Sci. Technol.*, **16**, (2), 600 (1979).

tremely valuable in such studies, yielding information that could not be obtained by any other means. Its use will undoubtedly increase because of the recent advent of sophisticated computer-interfaced data-collection systems which are being developed for rapid-scan wavelength-kinetic experiments.

The investigations of Ainsworth[34,35] paved the way to applying factor analysis to kinetics. In his pioneering work, Ainsworth[34] used the rank-analysis method of Wallace[1] to determine the number of absorbing species in reaction mixtures. A venom-solubilized preparation of cytochrome oxidase was catalytically reduced and the reduced cytochrome oxidase was reacted with oxygen. Absorbances were measured at 11 wavelengths at four specific time intervals using a stopped-flow technique. The rank of the data matrix was found to be three, implying that at least three components were present. This was interpreted as the result of two successive reactions of the type A → B → C.

Ainsworth[34] also studied reaction mixtures of oxyhemoglobin and reduced hemoglobin. Because the rate of reaction with oxygen was slow, no appreciable change in composition occurred during the time needed to obtain a set of absorbance readings of the reaction mixture. Rank analysis of the data gave evidence that four heams of the hemoglobin molecule were present.

In another study, Ainsworth[35] recognized that it was possible, under certain conditions, to obtain the absorption spectrum of a component in the mixture without prior knowledge of the identities or spectra of any of the components. This may be accomplished when there exists a situation where the component does not contribute to the total absorption. Often this condition exists at the very beginning or at the very end of a chemical reaction, when the concentration of either a product or reactant is negligibly small. When this occurs, the rank of the absorbance matrix is diminished by one unit. By calculating the determinants of all the submatrices, it is possible to determine whether or not such a situation exists. If it does, then a further study, at times when the component does contribute to the absorption, will reveal the spectrum of the component in question. Ainsworth successfully applied this technique to artifically computed data concerning mixtures of acridene orange, diiodo(R)-fluorescein, and rhodamine B in alcohol.

The studies of Wallace[1] and Ainsworth[34,35] were based on determining the rank by examining submatrices for singularity using the standard deviation criterion. As described in Section 7.1.2, Katakis[5] deduced the rank by examining the residual matrices produced by the Gauss process of elimination. Using this procedure, he investigated the reaction between Cr^{2+} and maleic acid in 1 M perchloric acid solution. His absorbance data matrix consisted of wavelengths as rows and reaction times as columns. The fact that the first residual matrix was found to be less than the error matrix showed conclusively that only one absorbing component was present. Such information was a valuable aid in deducing the true mechanism of the reaction.

In order to study the kinetics of the hydrolytic depolymerization of certain cobalt(III) complexes, Hugus and El-Awady[13] developed and applied a variety of factor analytical techniques (see Section 4.3). The results and conclusions of their studies are described in detail in Section 7.1.3.

Cochran and Horne[36] discussed the problems of applying factor analysis to wavelength-kinetic experiments. Using mathematical models, they showed that experimental errors that vary with wavelength can lead to incorrect estimates of the number of species. They then showed how this error can be eliminated by statistically weighting the absorbance matrix. This method is not restricted to wavelength kinetics but is applicable in a more general sense. Unfortunately, they did not apply it to any real kinetic problems.

7.4 OPTICAL ROTATORY DISPERSION

Rank analysis can be used to determine the number of optically active species in a mixture when the angle of rotation of a plane-polarized beam of light of each component is proportional to the concentration of the component. Under these conditions, an equation analogous to (7.2) can be written:

$$\alpha_{\lambda k} = \sum_{j=1}^{n_c} \bar{\alpha}_{\lambda j} c_{jk} \tag{7.10}$$

Here $\alpha_{\lambda k}$ is the angle of rotation of the kth mixture observed at wavelength λ, $\bar{\alpha}_{\lambda j}$ the specific rotation of the jth component observed at λ, c_{jk} the concentration of component j in mixture k, and n_c the number of optically active components in the mixtures. An optical rotatory dispersion (ORD) curve is obtained by measuring the angle of rotation as a function of wavelength.

McMullen et al.[37] applied the matrix rank analysis method of Wallace and Katz[2] to investigate the optical rotatory dispersion of tobacco mosaic virus TMV RNA in solution. Their purpose was to determine the number of components present and to identify them.

The ORD of TMV RNA was measured in the wavelength region between 230 and 350 nm, over a wide range of temperatures and at four different ionic strengths. The resultant data for a specific ionic strength was digitized into a matrix in which the row designees were the different wavelengths and the column designees were the different temperatures. Four data matrices were examined, one for each ionic strength. This methodology was quite unique because temperature was used to alter the compositions. It was based on the premise that the equilibrium between different geometrical conformers would be temperature-dependent. The method was fruitful; in each case, rank analysis revealed that two components were present. An attempt was made to fit all the experimental spectra with only two typical spectra chosen from the experimental data itself. The low-temperature spectrum of the 1 M Na$^+$ case was chosen to represent the first vector. The second vector was chosen to be the high-temperature spectrum of the 0.004 M Na$^+$ case. These typical vectors were selected because they represent two extremes, particularly so since increasing ionic strength had the same effect on the spectrum as decreasing the temperature. With these two typical vectors the entire data could be reproduced to within 4%.

Encouraged by these results, McMullen and coworkers attempted to transform the two factors into two different forms of TMV RNA molecule. They postulated that these two forms were the single-strand and double-strand helical conformations which coexist in chemical equilibrium. At high temperatures the equilibrium is shifted so that very little double strand remains. They proposed that the optical rotation of the single strand depends upon temperature, whereas that of the double strand is independent of temperature. The optical rotation of both conformers are insensitive to ionic strength. However, a change in ionic strength shifts the equilibrium. The model led to a direct calculation of the percentage composition of the double strand and the equilibrium constant within the range of the experimental conditions. At 2.5°C, TMV RNA appears to be approximately 50% double helix, whereas at 74.5°C it appears to be approximately 3% double helix.

7.5 CHROMATOGRAPHIC RESOLUTION

MacNaughtan et al.[38] were the first to report the successful application of factor analysis for deconvoluting two or more overlapping peaks in chromatography. The method requires several chromatograms of mixtures having different compositions but the same components. High precision, particularly on the time axis, is necessary. The data matrix is constructed by digitizing each chromatogram at equal intervals of time. Each row of the data matrix corresponds to a given mixture and each column corresponds to a given elution time.

One of their studies was concerned with mixtures of benzene and perdeuterobenzene. The chromatograms of four mixtures were recorded. The areas were normalized to unity to compensate for any errors due to variations in sample size. The resulting data was factor-analyzed and subjected to a deconvolution program. The quantitative results of the deconvolution study agreed within 2% of the results obtained by complete chromatographic separation.

The deconvolution program has one limitation: the chromatogram must have regions arising from each of the pure components. For a two-component system, this restriction is not too severe because the two extreme tails of the chromatogram meet this criterion.

Davis and coworkers[39] made use of mass spectra to determine the number of components under a single chromatographic peak. They recorded the mass spectra at fixed time intervals during the elution of a chromatographic peak. Thus each scan recorded the complete mass spectrum of a different composition of the same components. The data matrix consisted of the mass spectral intensities, wherein each column designated a time interval and each row designated a given mass-to-charge ratio. This data matrix was then factor-analyzed.

The method was successfully applied to isotopic mixtures of carbon dioxide, $^{13}C^{16}O_2$ and $^{12}C^{16}O_2$, and to mixtures of *n*-hexane and *n*-heptane. The investigation also included a study of the effects of differences in chromatographic resolution, peak heights, peak widths, and peak tailing. Peak distortion from chemical or electronic sources, channel-to-channel carryover and changes in base line were found to have no significant effect on the ability of FA to detect the second component. Noise constituted the most serious problem, sometimes producing a ficticious component. However, this situation could be diagnosed quickly by visual inspection of the experimental graphs.

Since the entire PFA calculations can be carried out in 3 to 5 minutes for a 200×5 matrix, this approach affords a quick and useful method for detecting the presence of more than one component in a chromatographic peak which may appear to be due to a single species. In contrast to the usual deconvolution technique, the factor analysis method requires no prior assumption concerning the chromatographic peak shape of any component. The method can thus be used to confirm the purity of a peak or to give warning that the chromatographic separation has not been effective. A minicomputer interfaced with a GC–MS system is especially appropriate for such studies. The FA method is rapid, sensitive, and reliable.

7.6 MASS SPECTRA

7.6.1 Abstract FA Studies

The use of FA–MS (factor anslysis of mass spectra) for component analysis was explored by Ritter et al.[40] Mass spectral data of mixtures are factor-analyzable because the signal intensity at a given m/e position is a linear sum of the corresponding intensities of the pure components weighted by their compositions. Ritter and coworkers factor-analyzed the mass spectra of four sets of mixtures: cyclohexane/cyclohexene, hexane/cyclohexane, heptane/octane, and unknown xylenes. The mixtures in a given set contained the same two components but differed in their composition. The data matrix consisted of the MS intensities measured at the same m/e positions for each mixture belonging to the given set. By studying the residual error in the covariance matrix (see Section 4.2.7), these investigators correctly identified the number of components in the mixtures and predicted that the unknown xylenes contained three components.

Not all m/e positions need be recorded in the data matrix that is to be factor-analyzed. In fact, by deleting particular m/e positions from the data matrix, one can effectively identify components that have unique mass positions. For example, deleting m/e 28 from the hexane/cyclohexane data matrix, Ritter and coworkers found that the factor space was reduced from 3 to 2, giving clear evidence that nitrogen was present as an impurity. Factor analysis of the heptane/octane mixtures indicated that three components instead of two were present. Careful examination of the spectra showed that the ion source was contaminated by nitrobenzene derivatives which had been run earlier.

This sutdy showed that FA–MS can be used as a rapid and accurate method of determining the number of components in a series of mixtures.

7.6.2 Target Factor Analysis Studies

Malinowski and McCue[41] showed how target factor analysis of mass spectral data could be used for qualitative identification of substances suspected to be present in a series of related mixtures. They also showed how TFA can be used to obtain the chemical compositions of the mixtures as well. This unique approach to compound identification and subsequent quantitative analysis illustrates the power of TFA in analytical chemistry.

The basis of the methodology is as follows. The intensity (height) of each mass peak in the MS of a mixture is a linear sum of contributions due to each component:

$$H(i, \alpha) = \sum_{j=1}^{n} h^0(i, j)p(j, \alpha) \qquad (7.11)$$

where $H(i, \alpha)$ is the height of the ith m/e peak in mixture α, $h^0(i, j)$ the height of the ith peak of the pure jth component per unit pressure, and $p(j, \alpha)$ the partial pressure of the component in the ionization chamber. Because of mass discrimination, the ratio of the partial pressures in the ionization chamber to that in the sample reservoir is different for each component. Because these pressures are extremely low, Dalton's law applies, so that

$$H(i, \alpha) = \sum_{j=1}^{n} H^0(i, j)F(j, \alpha) \qquad (7.12)$$

where

$$F(j, \alpha) = X(j, \alpha) \frac{D(j)}{p^0(j)} P(\alpha) \qquad (7.13)$$

Here $H^0(i, j)$ is the height of the ith m/e peak in the MS of pure j at pressure $p^0(j)$ in the ionization chamber, $P(\alpha)$ the total pressure in the ionization chamber, $D(j)$ the mass discrimination factor, and $X(j, \alpha)$ the mole fraction of j in the original sample mixture α.

Equation (7.12) shows that the spectral heights of the pure components are true factors and can be used as test vectors in TFA. Equation (7.13) shows how the corresponding cofactors, $F(j, \alpha)$, are related to the mole fractions. By employing a solution of known composition, one can obtain the compositions of the mixtures, independent of pressure measurements. This is possible because, for a given solution having components 1, 2, . . . , n, the ratios of the $F(j, \alpha)$ cofactors are independent of the total pressure $P(\alpha)$.

In order to use FA–MS for qualitative and quantitative analysis of mixtures, the following sequence of operations is carried out. First the number of components is deduced by decomposing the covariance matrix. Then the components are identified by using the MS of the pure components, suspected to be present, as test vectors in TFA. Finally, the compositions are obtained by adding the MS of a solution of known composition to the data matrix and then carrying out combination TFA using the MS of the pure components as real vectors.

In order to illustrate the steps in the TFA technique, Malinowski and McCue[41] subjected the MS data of Ritter and coworkers[40] to the foregoing sequence of operations. The data matrix consisted of the intensities of 18 m/e values for 7 cyclohexane/hexane mixtures. A previous AFA study[14] of this data matrix gave the results shown in Table 7.3. Evidence for the presence of three components was given by the fact that the IE function showed little or no improvement on using four or more eigenvectors and the fact that the IND function reached a minimum at $n = 3$. The third component was suspected to be nitrogen gas as a contaminant.[40] When the intensities of the mass 28 peak, characteristic of nitrogen, were deleted from the data matrix the results given in Table 7.3 were obtained. Both the IE and IND functions showed that only two components were

Table 7.3   Factor analysis results concerning mass spectra intensities used to determine the number of components in a series of related mixtures[a]

	Cyclohexane/Hexane			Cyclohexane/Hexane (Without m/e 28)			
n	RE	IE	IND × 10³	n	RE	IE	IND × 10³
1	1.810	0.684	50.27	1	1.812	0.685	50.35
2	0.465	0.249	18.62	2	0.134	0.071	5.36
3	0.128	0.084	8.03	3	0.106	0.070	6.65
4	0.111	0.084	12.30	4	0.092	0.070	10.25
5	0.098	0.073	24.56	5	0.072	0.061	18.08
6	0.074	0.068	73.51	6	0.058	0.054	58.18

[a] Reprinted with permission from E. Malinowski, *Anal. Chem.*, **49**, 612 (1977).

responsible for the remaining spectral data, thus confirming the presence of nitrogen. This was the same conclusion reached by Ritter and coworkers, who used the residual error in the covariance matrix as the error criterion. From Table 7.3 we see that the real error (RE) corresponding to $n = 3$ for the total data matrix and corresponding to $n = 2$ for the resuced data matrix (without m/e 28) is 0.13 intensity unit. This is considerably greater than the error, ±0.05, reported by the original investigators. The value 0.13 is much more reliable since it is a composite of all sources of error, whereas the value 0.05 was simply the error in reading the MS intensities from the experimental graphs.

Using the reduced data matrix and two factors, target tests were carried out[41] using the MS intensities of pure cyclohexane and pure hexane as test vectors. In both cases the predicted intensities agreed with the test vectors within the expected error limit 0.13, as shown in Table 7.4 for cyclohexane. Intensities for those masses which were free-floated in the test vector were predicted correctly, thus providing further evidence for the presence of these components.

When the two test vectors were used in combination, TFA yielded the complete set of $F(j,\alpha)$ cofactors. Mixture 4, containing 55 mol % cyclohexane, was considered to be the standard solution of known composition. Using (7.13), the values of the $F(j, \alpha)$ cofactors and the known composition of mixture 4, Malinowski and McCue[41] determined the compositions of the solutions given in Table 7.5. The agreement between the calculated and reported compositions is good considering the fact that the original solutions were prepared crudely.

In an elaborate study involving target testing, Rasmussen et al.[42] compiled a library file of approximately 17,000 mass spectra, which was used to correctly identify the components in several different mixtures and to estimate their relative concentrations. Their computational strategy was made efficient by employing a prefilter to remove library entries having masses greater than the

Table 7.4   Mass spectral intensities of cyclohexane obtained from target testing[a] and from spectral isolation[b]

m/e	Test[c]	TFA Prediction	Spectral[d] Isolation
27	(1.8)	1.9	1.8
29	1.3	1.3	1.1
39	2.5	2.3	2.3
40	0.7	0.6	0.7
41	(7.1)	7.3	6.9
42	(3.5)	3.5	3.2
43	2.2	2.1	1.8
44	0.2	0.2	0.1
54	(0.8)	0.7	0.8
55	4.6	4.9	4.7
56	13.5	13.6	13.5
57	(1.2)	1.2	0.7
69	3.8	4.1	4.0
83	0.8	0.7	0.7
84	10.7	10.4	10.7
85	0.9	0.9	0.9
86	(0.1)	0.1	0.0

[a] Reprinted with permission from E. R. Malinowski and M. McCue, *Anal. Chem.*, **49**, 284 (1977).

[b] Reprinted with permission from F. J. Knorr and J. H. Futnell, *Anal. Chem.*, **51**, 236 (1979).

[c] Values in parentheses were free-floated (i.e., left blank in the test vector).

[d] Adjusted so that the base peak is 13.5 rather than 100, as reported in the original paper.

highest mass of the mixture spectra. The validity of a target was judged by the following approach, which made use of Bessel's inequality test. The *coefficient of fit, b*, was calculated by taking the sum of the squares of the dot products between the normalized test vector and each column of the normalized row-factor matrix. This sum can be shown to be equal to the following:

$$b = \left\{ \frac{\overline{\overline{R}}_l [R^{\ddagger}][\lambda^{\ddagger}]^{-1/2}}{(\overline{\overline{R}}_l \cdot \overline{\overline{R}}_l)^{1/2}} \right\}^2 \tag{7.14}$$

The coefficient of fit measures the extent to which the target vector lies in the factor space. According to Bessel's inequality, a value of 0 indicates that the target is completely orthogonal to the factor space and lies entirely outside the space. A value of 1 indicates that the target lies completely inside the space.

Table 7.5   Composition of cyclohexane/hexane mixtures obtained from target factor analysis[a] and from spectral isolation[b]

| Mixture | Experiment | Mole Fraction Cyclohexane | |
		TFA (Prediction)	Spectral Isolation (Prediction)
1	1.00	1.00	0.96
2	0.92	0.88	0.84
3	0.83	0.81	0.78
4	0.55[c]	0.55[c]	0.54
5	0.23	0.30	0.30
6	0.12	0.16	0.17
7	0.00	0.00	0.01

[a] Reprinted with permission from E. R. Malinowski and M. McCue, *Anal. Chem.,* **49,** 284 (1977).

[b] Reprinted with permission from F. J. Knorr and J. H. Futrell, *Anal. Chem.,* **51,** 236 (1979).

[c] Represents the standard solution.

Hence target vectors associated with true components will have *b* values close to unity. This approach provides us with a method of searching through a huge library of spectra and sorting out the true components of mixtures.

Under certain conditions, the mass spectra of the components can be separated from the mass spectra of mixtures without recourse to library information. This can be achieved if there exists at least one mass peak in the mixed spectra which is unique to each component. Such pure mass peaks are most likely to occur when the number of components is small and the number of mass points is large and divergent. The pure mass points need not be specified by the chemist; instead, they are automatically selected by the algorithm involved.

A method for selecting the pure mass points, similar to the separation technique of Lawton and Sylvestre,[23] was developed by Knorr and Futrell.[43] First, the row-factor matrix $[R^{\ddagger}]$ is normalized so that the sum of the squares across each row is unity. This places each mass point on the surface of an n-dimensional sphere subtended by real, oblique axes which pass through the pure mass points. The first pure mass is identified by finding the mass point that has the smallest contribution on the first eigenvector (i.e., the point having the smallest normalized r_{i1}). The second pure mass is chosen to be the one whose contribution on the second eigenvector (normalized r_{i2}) differs most significantly from the second eigenvector contribution of the first pure mass point. The third pure mass is selected by finding the mass whose difference in the third eigenvector contribution is greatest between itself and the average of the third contributions of the previously chosen pure masses. To search for the remaining pure masses, this process of finding the largest difference between the successive, normalized eigenvector contributions of the newly sought mass and the average of the previously chosen masses is continued until all n pure mass points have been established. The normalized row-cofactor contributions for the pure masses specify the relative amounts of the eigenvectors that comprise those mass points. The n rows of normalized row cofactors, associated with the n pure masses, constitute the rows of the inverse of the transformation matrix, $[T]^{-1}$. Having, thus found the transformation matrix, the abstract factor matrices can be converted into real matrices, one that traces out the mass spectra of the pure components and the other, which reveals the relative concentrations of the components.

When this technique was applied[43] to the mass spectral data of Ritter et al.,[40] the spectra of the pure components and their relative concentrations were obtained, independent of any other information. A typical example of such spectral isolation is shown in Table 7.4, giving clear evidence for the presence of cyclohexane. Prediction of the compositions of the cyclohexane/hexane mixtures is given in Table 7.5.

REFERENCES

1. R. M. Wallace, *J. Phys. Chem.,* **64,** 899 (1960).
2. R. M. Wallace and S. M. Katz, *J. Phys. Chem.,* **68,** 3890 (1964).
3. S. Perlis, *Theory of Matrices,* Addison-Wesley, Reading, Mass., 1952, p. 45.
4. L. P. Varga and F. C. Veatch, *Anal. Chem.,* **39,** 1101 (1967).
5. D. Katakis, *Anal. Chem.,* **37,** 876 (1965).
6. E. Bodewig, *Matrix Calculus,* North-Holland, Amsterdam, 1959.
7. J. S. Coleman, L. P. Varga, and S. H. Mastin, *Inorg. Chem.,* **9,** 1015 (1970).
8. J. J. Kankare, *Anal. Chem.,* **42,** 1322 (1970).
9. J. L. Simmonds, *Photogr. Sci. Eng.,* **2,** 205 (1958).
10. J. L. Simmonds, *J. Opt. Soc. Am.,* **53,** 968 (1963).
11. R. L. Reeves, *J. Am. Chem. Soc.,* **88,** 2240 (1966).
12. G. Wernimont, *Anal. Chem.,* **39,** 554 (1967).
13. Z. Z. Hugus, Jr., and A. A. El-Awady, *J. Phys. Chem.,* **75,** 2954 (1971).
14. E. R. Malinowski, *Anal. Chem.,* **49,** 612 (1977).
15. E. R. Malinowski, in B. R. Kowalski (Ed.), *Chemometrics: Theory and Application,* ACS Symp. Ser. 52, American Chemical Society, Washington, D.C., 1977, Chap. 3.
16. G. T. Rasmussen, T. L. Isenhour, S. R. Lowry, and G. L. Ritter, *Anal. Chim. Acta,* **103,** 213 (1978).
17. E. R. Malinowski and M. McCue, unpublished work.
18. M. K. Antoon, L. D'esposito, and J. L. Koenig, *Appl. Spectrosc.,* **33,** 351 (1979).
19. J. T. Bulmer and H. F. Shurvell, *J. Phys. Chem.,* **77,** 256 (1973).
20. J. T. Bulmer and H. F. Shurvell, *Can. J. Chem.,* **53,** 1251 (1975).
21. J. T. Bulmer and H. F. Shurvell, *J. Phys. Chem.,* **77,** 2085 (1973).
22. J. Korppi-Tommola and H. F. Shurvell, *Can. J. Chem.,* **56,** 2959 (1978).

23. W. H. Lawton and E. A. Sylvestre, *Technometrics,* **13,** 617 (1971).
24. N. Ohta, *Anal. Chem.,* **45,** 553 (1973).
25. E. A. Sylvestre, W. H. Lawton, and M. S. Maggio, *Technometrics,* **16,** 353 (1973).
26. K. Yamaoka and M. Takatsuki, *Bull. Chem. Soc. Jap.,* **51,** 3182 (1978).
27. C. H. Lin and S. C. Liu, *J. Chin. Chem. Soc.,* **25,** 167 (1978).
28. C. H. Lin and L. C. Lin, *Proc. Natl. Sci. Counc. Republic of China,* **3,** 1 (1979).
29. T. Jarv, J. T. Bulmer, and D. E. Irish, *J. Phys. Chem.,* **81,** 649 (1977).
30. H. F. Shurvell and A. Dunham, *Can. J. Spectrosc.,* **23,** 160 (1978).
31. I. M. Warner, G. D. Christian, E. R. Davidson, and J. B. Callis, *Anal. Chem.,* **49,** 564 (1977).
32. C. N. Ho, G. D. Christian, and E. R. Davidson, *Anal. Chem.,* **50,** 1108 (1978).
33. S. W. Gaarenstroom, *J. Vac. Sci. Technol.,* **16** (2), 600 (1979).
34. S. Ainsworth, *J. Phys. Chem.,* **65,** 1968 (1961).
35. S. Ainsworth, *J. Phys. Chem.,* **67,** 1613 (1963).
36. R. N. Cochran and F. H. Horne, *Anal. Chem.,* **49,** 846 (1977).
37. D. W. McMullen, S. R. Jaskunas, and I. Tinoco, Jr., *Biopolymers,* **5,** 589 (1967).
38. D. MacNaughtan, Jr., L. B. Rogers, and G. Wernimont, *Anal. Chem.,* **44,** 1421 (1972).
39. J. E. Davis, A. Shepard, N. Stanford, and L. B. Rogers, *Anal. Chem.,* **46,** 821 (1974).
40. G. L. Ritter, S. R. Lowry, T. L. Isenhour, and C. L. Wilkins, *Anal. Chem.,* **48,** 591 (1976).
41. E. R. Malinowski and M. McCue, *Anal. Chem.,* **49,** 284 (1977).
42. G. T. Rasmussen, B. A. Horne, R. C. Wieboldt, and T. L. Isenhour, *Anal. Chim. Acta,* **112,** 151 (1979).
43. F. J. Knorr and J. H. Futrell, *Anal. Chem.,* **51,** 236 (1979).

8 Nuclear
Magnetic Resonance

This chapter concerns factor analytical studies of the effects of substituents and solvents on nuclear magnetic resonance (NMR) chemical shifts.

8.1 PROTON SOLVENT SHIFTS

NMR spectral features (chemical shifts and coupling constants) are strongly influenced by the solvent. This is unfortunate because it makes spectral interpretation difficult, but fortunate because it provides the chemist with a clue for probing the liquid solution state and for studying intermolecular interactions.

Homer[1] has reviewed the factors that are believed to contribute to the solvent shift. Theoretical expressions for many of these factors have been derived and attempts have been made to isolate one factor from another by the judicious choice of solvent, solute, and other experimental variables. All such attempts have had only limited success, chiefly because it is impossible to find experimental conditions such that the effects of all but one factor are held constant.

The series of TFA studies of Weiner and Malinowski[2-5] led to a complete solution of the proton solvent shifts of some simple nonpolar solutes. In this section we will systematically trace through the intricate steps which eventually lead to the complete solution. Details of the development are presented in this section from a chronological viewpoint in order to emphasize the systematic and deductive reasoning involved.

8.1.1 Key Solvents

The following is a brief description of the first investigation of proton NMR solvent shifts using TFA, reported by Weiner et al.[2] The goal of the investigation was to develop a procedure for predicting the shifts of simple solutes in a large variety of solvents using a minimum of shift data.

Proton shifts of a series of simple substituted (polar and nonpolar) methanes, measured in a variety of solvents with tetramethylsilane (TMS) as an internal standard, were chosen as an ideal system to study. The raw data that were subjected to factor analysis consisted of the proton shifts of nine halogenated methanes in nine solvents. The shifts of CH_4, CH_3CN, CH_2Cl_2, CH_2ClCN, and $CHBrCl_2$ were not included in the data matrix but were purposely set aside for later testing purposes. The factor analytical reproduction step showed that three factors were sufficient to span the factor space and to reproduce the data within experimental error, ± 0.5 Hz. This implied that only three solvent–solute interaction terms were involved.

Since the nature of the interaction terms were unknown, attempts were made to find a set of "typical" factors. A typical factor is a column of the data matrix. Although any column of data can be used as an axis of the factor space, care must be exercised in choosing a set of three data columns, since an arbitrary combination of three columns may not span the factor space. A given combination may contain data that do not involve a particular solute–solvent interaction. For example, if hydrogen bonding is an important interaction, it is imperative that at least one of the three axes account for this factor.

Acetonitrile, carbon tetrachloride, and methylene bromide were chosen as a key set of typical solvent vectors for the following reasons. Acetonitrile possesses a large dipole moment and has pi electrons. Methylene bromide has a large polarizability and a sizable quadrupole moment. Carbon tetrachloride is nonpolar and contains bulky chlorine atoms. It was believed that these features would adequately account for all possible solute–solvent interactions involved in the factor space. In fact, after all combinations of typical column vectors were tar-get-tested, these three solvent factors were found to give the best data repro-duction. Although there is nothing unique about this choice, other combinations did not span the factor space. For example, the set composed of methylene chloride, chloroform, and carbon tetrachloride did not satisfactorily reproduce the data. Evidently, at least one of the important solute–solvent factors of the space was not sufficiently represented by this group.

A simultaneous combination transformation onto the three data columns yielded equations of the following kind:

$$\delta(u, CH_3CN) = 1.002f_1 \quad - 0.004f_2 + 0.002f_3$$
$$\delta(u, CH_2Cl_2) = 0.081f_1 \quad + 0.715f_2 - 0.207f_3$$
$$\delta(u, CHCl_3) = -0.046f_1 + 0.817f_2 + 0.230f_3$$
$$\delta(u, CCl_4) = -0.002f_1 + 1.004f_2 - 0.002f_3 \qquad (8.1)$$
$$\delta(u, CS_2) = 0.006f_1 \quad + 1.128f_2 - 0.139f_3$$
$$\delta(u, CH_3I) = 0.561f_1 \quad - 0.224f_2 + 0.653f_3$$

Here $\delta(u, v)$ is the chemical shift of solute u in solvent v and $f_1 = \delta(u, CH_3CN)$, $f_2 = \delta(u, CCl_4)$, and $f_3 = \delta(u, CH_2Br_2)$. These equations predict the chemical shift of a solute in a given solvent in terms of its measured shift in the three key solvents. Because of experimental error and computer roundoff, the equations given above for acetonitrile, methylene bromide, and carbon tetrachloride each exhibit three finite coefficients, two near zero and one near unity, rather than one.

The utility, as well as validity, of these equations was further substantiated by examining shifts predicted for the solutes which were purposely omitted from the factor analysis scheme. For example, by measuring the shifts of CH_2Cl_2 in the three key solvents, we can use (8.1) to predict the shift of methylene chloride in the other solvents. The shifts of CH_2Cl_2 in the three key solvent (CH_3CN, CCl_4, and CH_2Br_2) are, respectively, 326.9, 317.1, and 321.2 Hz. Placing these values for f_1, f_2, and f_3 into the (8.1) expression corresponding to CH_3I solvent, we make the following calculation:

$$\delta(CH_2Cl_2, CH_3I) = (0.561)(326.9) - (0.224)(317.1)$$
$$+ (0.653)(321.2) = 322.2 \text{ Hz}$$

This value, representing the proton shift of CH_2Cl_2 in CH_3I solvent, agrees with the measured value, 322.5 Hz, within experimental error, ±0.5 Hz.

Predictions for the methanes not included in the analysis were also found to be in excellent agreement with their measured shifts (see Table 8.1). Equations (8.1) have also been used to predict the shifts of substituted ethanes.[2] This was accomplished by measuring the shifts of various ethanes in the three key solvents. Typical predicted shifts for CH_2ClCCl_3 are shown in Table 8.1. We see here

Table 8.1 Experimental and predicted shifts^a of some new solutes using (8.1)^b

	Solute					
	CH$_2$ClCN		CHClBr$_2$		CH$_2$ClCCl$_3$	
Solvent	Experimental	Predicted	Experimental	Predicted	Experimental	Predicted
CS$_2$	242.8	241.9	427.9	427.5	254.1	253.8
CHCl$_3$	246.1	245.7	432.1	431.0	257.2	256.8
CHBr$_3$	253.0	252.3	433.0	433.0	—	261.1
CH$_3$I	255.3	254.2	439.6	440.4	263.0	264.6
CH$_2$I$_2$	257.3	258.5	434.9	434.9	—	241.8

^{a} In hertz at 60 MHz, relative to internal TMS.
^{b} Reprinted with permission from P. H. Weiner, E. R. Malinowski, and A. R. Levinstone, *J. Phys. Chem.*, **74**, 4537 (1970).

that even though the real nature of the factors is unknown, it is still possible to predict solvent shifts by means of factor analysis.

8.1.2 Theoretical Considerations

Encouraged by the success of the initial factor analytical studies, Weiner et al.[2] pursued the problem in a more fundamental way, reasoning as follows: Without proof, Buckingham et al.[6] postulated that the solvent shift can be expressed as a linear sum of terms:

$$\delta(u, v) = \delta(u, \text{gas}) + \sigma_b(v) + \sigma_a(v) + \sigma_w(u, v) + \sigma_E(u, v)$$
$$+ \sigma_H(u, v) + \cdots \quad (8.2)$$

where $\delta(u, v)$ is the chemical shift of solute u in solvent, v, $\delta(u, \text{gas})$ is the gas-phase shift of the solute, $\sigma_b(v)$ is due to the bulk susceptibility of the solvent, $\sigma_a(v)$ is the solvent shift caused by the anisotropy of the solvent, $\sigma_w(u, v)$ is the van der Waals dispersion interaction between the solute and solvent, $\sigma_E(u, v)$ is the reaction field interaction between the solute and solvent, and $\sigma_H(u, v)$ is due to hydrogen bonding.

Factor analysis gave evidence that the solvent shift must be a linear sum and that each of the terms must be a product function of solute and solvent parameters. Furthermore, the hydrogen-bonding contribution should be negligible, since none of the solutes and solvents studied form strong hydrogen bonds. Hence (8.2) can be written as

$$\delta(u, v) = \delta(u, \text{gas}) \cdot 1 + 1 \cdot \sigma_b(v) + 1 \cdot \sigma_a(v) + \sigma_w(u) \cdot \sigma_w(v)$$
$$+ \sigma_E(u) \cdot \sigma_E(v) \quad (8.3)$$

Equation (8.3) involves five factors, whereas factor analysis clearly indicated that only three factors are operative. The answer to this dilemma lies in the fact that the chemical shifts were referenced with respect to a trace of TMS which was dissolved in the same solvent. The internal standard, TMS, also comes in contract with the solvent and experiences a solvent shift:

$$\delta(\text{TMS}, v) = \delta(\text{TMS, gas}) \cdot 1 + 1 \cdot \sigma_b(v) + 1 \cdot \sigma_a(v) + \sigma_w(\text{TMS}) \cdot \sigma_w(v)$$
$$+ \sigma_E(\text{TMS}) \cdot \sigma_E(v) \quad (8.4)$$

The experimental shift, $\delta^{\text{TMS}}(u, v)$, represents, in reality, the difference between (8.3) and (8.4):

$$\delta^{\text{TMS}}(u, v) = \delta^{\text{TMS},g}(u, \text{gas}) \cdot 1 + [\sigma_w(u) - \sigma_w(\text{TMS})]\sigma_w(v)$$
$$+ [\sigma_E(u) - \sigma_E(\text{TMS})]\sigma_E(v) \quad (8.5)$$

where $\delta^{\text{TMS},g}(u, \text{gas}) = \delta(u, \text{gas}) - \delta(\text{TMS, gas})$ is the gas-phase shift of the solute relative to the gas-phase shift of TMS. Equation (8.5) involves a sum of three terms. Each term is a product of a solute and solvent contribution. This expression predicts that the data space is three-dimensional, in complete accord with the results of factor analysis.

8.1.3   Solute Gas-Phase Shift

According to (8.5) the gas-phase shift of the solute should be a basic factor. Because each suspected factor can be examined independently, via target transformation, the gas-phase shift can be tested solely on its own merits, without our having to invoke any model or specify any details concerning the other two factors. Transformation into the solute gas-phase shifts, relative to gaseous TMS, was successful (see Table 8.2), giving clear evidence that the gas-phase shift is a true fundamental factor.

In Chapter 2 we learned that not all the data points are required for target transformation. A fringe benefit of TFA is its ability to predict those points not inserted (free-floated) in the target test vector. In Table 8.2 the second column shows predicted gas-phase shifts for CH_3Cl, CH_2I_2, CHI_3, and CH_2ClBr which were free-floated. In spite of the fact that two other unknown factors are simultaneously active, we see how valuable information can be obtained by means of a single successful target transformation.

Target transformation into gas-phase shifts relative to methane gas failed. The shifts involved in the test vector must be referenced with respect to gaseous TMS as dictated by the theoretical equation (8.5). The choice of a reference standard in the test vector is not arbitrary. This illustrates the sensitivity of the target test to the exact details of the basic factor being investigated.

Table 8.2   Test of gas-phase chemical shifts[a] as a solute factor using three factors[b]

Solute	Test	Predicted
CH_3Cl	—	168.2
$CHCl_3$	427.3	427.1
CH_3Br	146.9	147.1
CH_2Br_2	285.0	285.5
$CHBr_3$	406.9	406.8
CH_3I	119.0	118.5
CH_2I_2	—	227.6
CHI_3	—	301.5
CH_2ClBr	—	297.7

[a] In hertz at 60 MHz, relative to gaseous TMS.
[b] Reprinted with permission from P. H. Weiner, E. R. Malinowski, and A. R. Levinstone, *J. Phys. Chem.*, **74**, 4537 (1970).

8.1.4   External Standards

Subsequent discussions require an understanding of NMR referencing procedures. There are two different methods for measuring and reporting solute chemical shifts: internal and external reference standards.

The internal standard method, the more common procedure, involves dissolving a trace of a reference compound in the solute–solvent solution and observing the shift between the reference and the solute. There are several advantages to this procedure. The main experimental advantage is the ease of measurement. This procedure also has a theoretical advantage in that the bulk magnetic susceptibility term, $\sigma_b(v)$, and the solvent anisotropy term, $\sigma_a(v)$, are eliminated [see (8.5)]. These two terms depend primarily upon the solvent and produce the same shift effect on both the solute and the internal reference. The primary disadvantage of using an internal standard is that one measures not the effect of the solvent on the solute, but the difference between the effect of the solvent on the solute and the effect of the solvent on the reference compound. Also, anomolous intermolecular interactions between the reference and solute molecules may occur, seriously complicating the situation.

The second major experimental approach for measuring solvent effects involves the use of an external standard. In this case the reference material and solution are placed in separate compartments of a coaxial cell consisting of two concentric glass cylinders. An external reference is a reliable standard because it does not come in contact with the solute solution. However, since the reference standard and the solute do not experience the same bulk magnetic susceptibility shielding effect, the observed chemical shift must be appropriately corrected. Theoretical equations involved in this correction are well established. In fact, an NMR technique[7] can be used to determine the bulk susceptibility corrections. In addition to making the problem more complex from an experimental point of view, the use of external standards allows the solvent anisotropy effect to be active. Although this increases the complexity of the theoretical problem, more useful information can be obtained.

8.1.5 Solvent Anisotropy

The success of the earlier investigation[2] encouraged Weiner and Malinowski[3] to probe deeper into the theoretical aspects of the problem. They decided to simplify the problem by studying a subspace composed of nonpolar solutes. In this situation the reaction field, $\sigma_E(u, v)$, is absent because nonpolar solutes lack a permanent electric dipole moment necessary to produce a reaction field. Hexamethyldisiloxane (HMD) was used as an external standard and corrections were made for bulk susceptibility, $\sigma_b(v)$. The proton shifts of six nonpolar solutes (methane, ethane, neopentane, cyclohexane, cyclooctane, and tetramethylsilane) dissolved in 22 solvents were measured. The experimental accuracy of the data, knowledge crucial to proper factor analysis, was estimated to be ± 1.5 Hz.

When the shift data were initially factor-analyzed, two factors almost seemed sufficient to reproduce the data matrix within experimental error. Only a small improvement was observed when three factors were invoked. With a two-factor model, one might erroneously conclude that the solvent anisotropy effect was negligible. However, the two-factor model was rejected in favor of a three-factor model when serious inconsistancies arose. The nature of these inconsistencies will become apparent at a later point in our discussion.

To study the three fundamental factors, we again examine the equation of Buckingham, Schaefer, and Schneider, which can be appropriately expressed as follows:

$$\delta^{HMD,X}(u, v) = \delta^{HMD,X}(u, \text{gas}) \cdot 1 + \sigma_w(u) \cdot \sigma_w(v) + 1 \cdot \sigma_a(v) \quad (8.6)$$

where $\delta^{HMD,X}(u, v)$ is the chemical shift of solute u in solvent v, relative to external HMD, and $\delta^{HMD,X}(u, \text{gas})$ is the gas-phase shift of solute u relative to external HMD (the shifts having been corrected for bulk susceptibility).

According to factor analysis, each term in this equation must be a product function of solute and solvent parameters. For the gas-phase term, this constraint is satisfied since the gas-phase chemical shift is only a function of the solute being studied, independent of the solvent. Similarly, the solvent anisotropy term is dependent entirely on the solvent and is independent of the solute. These two terms can therefore be expressed as product functions with the solvent part of the former term being unity and the solute part of the latter term equaling unity. Several investigators[8,9] have suggested that the van der Waals shielding can be expressed as a product function, in accord with (8.6).

To test this equation in detail, we need accurate estimations for all three terms on the right side of this equation. Such information is not readily available since the van der Waals term and solvent anisotropy are theoretical quantities. For these quantities, only crude estimates for a very limited number of solvents exist. Theoretical estimates of solvent anisotropy have been made for only two solvents, benzene and carbon disulfide. Schug[10] estimated the anisotropies of these solvents to be -30 Hz and $+18.1$ Hz, respectively, whereas Homer[11] suggested -35 Hz and $+7$ Hz, respectively.

This perplexing problem was solved by Weiner and Malinowski[3] in the following manner. For methane as a solute, (8.6) takes the form

$$\delta^{HMD,X}(CH_4, v) = \delta^{HMD,X}(CH_4, gas) \cdot 1 + \sigma_w(CH_4) \cdot \sigma_w(v)$$
$$+ 1 \cdot \sigma_a(v) \quad (8.7)$$

This equation was solved for $\sigma_w(\alpha)$ and substituted into (8.6), yielding

$$\delta^{HMD,X}(u, v) = \delta^{HMD,X}(u, gas) \cdot 1 + \frac{\sigma_w(u)}{\sigma_w(CH_4)} \cdot \delta^{CH_4,g}(CH_4, v)$$
$$+ \left[1 - \frac{\sigma_w(u)}{\sigma_w(CH_4)} \right] \cdot \sigma_a(v) \quad (8.8)$$

In this expression, $\delta^{CH_4,g}(CH_4, v)$ represents the gas-to-solution shift of methane in solvent v. The advantage of this expression in comparison to (8.6) lies in the fact that it is independent of the van der Waals model, circumventing the need to specify $\sigma_w(v)$. Instead, this solvent factor is replaced by a new factor, the methane gas-to-solution shift. This procedure illustrates the intricate interdependency between fundamental factors and the fact that there are many different ways of expressing the factor space, each way being an equally valid representation.

From the viewpoint of solvent tests, the unity and methane-gas-to-solution vectors are easily constructed since all necessary data are available. On the other hand, the solvent anisotropy test vector poses a serious problem, as described earlier. To solve this problem, Weiner and Malinowski[3] proceeded as follows. They ascribed a zero value for the solvent anisotropy of carbon tetrachloride because it is nonpolar and symmetric. They then sustematically varied the solvent anisotropies of benzene and carbon disulfide in accord with various theoretical estimates. Because there were three factors and only three solvents for which only crude estimates of solvent anisotropies were available, it was not possible to conclude which of these estimates were the best from the results of the target tests alone. This anomaly arose because, with a three-factor space, any test factor containing only three defined numbers would yield a perfect fit. Since, in this case, any three random numbers would fit perfectly, another criteria had to be devised.

Table 8.3   Solvent cofactor target tests[a]

Solvent	Unity		$\delta^{CH_4,\,g}_{(CH_4,\,v)}$		$\sigma_a(v)$		$\sigma_w(v)$	
	Test	Predicted	Test	Predicted	Test	Predicted	Test	Predicted
CCl_4	1.0	1.014	25.2	25.4	0.0	0.0	0.245	0.273
$CHCl_3$	1.0	1.009	23.0	23.1	—	0.8	0.238	0.242
$CHBr_3$	1.0	1.002	34.4	34.5	—	4.9	0.323	0.310
CH_2I_2	1.0	0.974	43.4	43.2	—	14.8	0.301	0.286
CH_2ClCCl_3	1.0	1.007	24.4	24.5	—	-0.1	—	0.264
C_6H_{12}	1.0	1.001	16.9	16.9	—	-1.3	—	0.203
CS_2	1.0	0.992	32.7	32.6	9.0	9.0	—	0.246
$(CH_3)_2CO$	1.0	1.003	9.5	9.5	—	-5.9	0.175	0.180
C_6H_6	1.0	0.977	-9.5	-9.7	-24.0	-24.0	0.199	0.186
C_6F_6	1.0	1.002	-12.5	-12.4	—	-27.6	—	0.200

[a] Reprinted with permission from P. H. Weiner and E. R. Malinowski, *J. Phys. Chem.*, **75**, 1207, 3160 (1971).

Equation (8.8) provides the necessary criteria for selecting the proper anisotropy test vector. According to this equation, the coefficients of the solvent unity test vector should correspond to the gas-phase chemical shifts of the solutes. Furthermore, the sum of the last two solute coefficients in (8.8) should equal unity. If a simultaneous target transformation into the three-solvent test vectors (unity, methane gas-to-solution shifts, and solvent anisotropy) is performed, the best anisotropy test vector can be deduced by comparing the predicted loadings [i.e., the solute coefficients in (8.8)] with the criteria given above.

The results of the target tests involving these three solvent factors are shown in Table 8.3. In order to obtain the best solvent anisotropy vector, a variety of test vectors were generated and examined in the following manner. The solvent anisotropy of carbon tetrachloride was set equal to zero, and all combinations of the estimated values of benzene and carbon disulfide were employed in the combination step along with the unity and methane gas-to-solution shifts test vectors. The solute coefficients generated by this procedure did not agree satisfactorily with the specified criteria.

A more detailed search was undertaken. The solvent anisotropy of carbon tetrachloride was again set equal to zero while the anisotropy value of benzene was systematically varied from -40 to -20 Hz and the value for carbon disulfide was varied from $+5$ to $+20$ Hz. For each combination, a comparison was made of the predicted solute coefficients with the proposed criteria. The best fit was obtained when σ_a(benzene) $= -24.0$ Hz and $\sigma_a(CS_2) = +9.0$ Hz, yielding the following equations:

$$\delta^{HMD,X}(CH_4, v) = -8.3f_1 + 0.99f_2 + 0.01f_3$$
$$\delta^{HMD,X}(CH_3CH_3, v) = 35.9f_1 + 0.72f_2 + 0.28f_3$$

$$\delta^{HMD,X}(\text{neo-}C_5H_{12}, v) = 41.6f_1 + 0.64f_2 + 0.32f_3$$
$$\delta^{HMD,X}(C_6H_{12}, v) = 75.8f_1 + 0.49f_2 + 0.55f_3$$
$$\delta^{HMD,X}(C_8H_{16}, v) = 83.8f_1 + 0.38f_2 + 0.63f_3$$
$$\delta^{HMD,X}(\text{TMS}, v) = 12.8f_1 + 0.63f_2 + 0.35f_3$$

Here f_1 = unity, f_2 = methane gas-to-solution shift, and f_3 = solvent anisotropy. The coefficients of the unity terms can be compared to the actual experimental gas-phase shifts[12] (relative to external HMD): methane = -8.4 Hz, ethane = 35.5 Hz, neopentane = 42.1 Hz, cyclohexane = 75.6 Hz, and tetramethylsilane = 16.4 Hz. Except for tetramethylsilane, the predicted gas-phase shifts shown in the equations above are in excellent agreement with the experimental values. Furthermore, for each equation shown, the sum of the solute coefficients for the last two terms on the right is close to unity. Since the two criteria concerning the solute coefficients have been satisfied, the solvent anisotropy test vector should be reliable.

The "best" anisotropy test vector is shown in Table 8.3. Solvent anisotropies predicted for the free-floated solvents are shown in the table. These predictions represent the first empirical estimations of solvent anisotropy. Halogenated solvents such as methylene iodide and bromoform are predicted to have large solvent anisotropies. These values serve as guides in the theoretical development of solvent anisotropies for the halogenated methanes.

Recall that the data matrix could be reproduced with only two abstract factors, but the two-factor solution was rejected in favor of a three-factor solution. This conclusion can now be understood in the light of the present discussion. With the two-factor model, the assumption is made that the anisotropy contribution to the chemical shift is negligible for all solvents in the scheme. This is a particularly poor assumption for solvents such as benzene and carbon disulfide. If this assumption is made, (8.8) reduces to

$$\delta^{HMD,X}(u, v) = \delta^{HMD,X}(u, \text{gas}) \cdot 1 + \frac{\sigma_w(u)}{\sigma_w(CH_4)} \cdot \delta^{CH_4,g}(CH_4, v) \quad (8.9)$$

When an equation of this form was tested, using a two-dimensional transformation matrix, the solute coefficients of the unity test differed from the true gas-phase values by more than 10 Hz. These large deviations were far beyond experimental error, approximately ± 1.5 Hz. The two-factor model was rejected in favor of the three-factor model, which, in marked contrast, did predict the gas-phase shifts within experimental error.

8.1.6   van der Waals' Effect

In Section 8.1.5 the van der Waals shift contribution was circumvented by using the experimentally determined methane gas-to-solution shifts as a measure of this term. This procedure yielded the "best" set of solvent anisotropies, independent of any particular theoretical model that might be proposed for the van der Waals contribution. Having obtained a satisfactory set of solvent anisotropies, Weiner and Malinowski[4] then attempted to unravel the van der Waals factor.

Any model of the van der Waals shift must meet four criteria. The first criteria concerns the overall ability of the three solvent test factors to predict the data matrix within experimental error. The last three criteria concern the solute coefficients in (8.6). If the van der Waals model is correct, the solute coefficients of the unity test factor should correspond to the solute gas-phase chemical shifts; the solute coefficients of the anisotropic shift term should be unity; and the solute coefficient of the van der Waals term should agree with the model.

Several theoretical models have been proposed to explain the van der Waals effect involving nonpolar solutes. Linder and coworkers[12,13] treated the solute as an oscillating dipole and the solvent as a dielectric continuum. Bernstein and coworkers[14] utilized a virial expansion of solute–solvent interactions. Both of these theories have much in common but differ in detail. Because of its relative simplicity, Weiner and Malinowski[3] tested the continuum approach of Linder. Following the theory of Linder, and after removing an unnecessary approximation, Weiner and Malinowski obtained the following expression for the van der Waals contribution:

$$\sigma_w(u, v) = \left[\frac{k}{V_u}\right]_u \left[\frac{n^2 + 2}{2n^2 + 1} \times \frac{\sum \langle r_j{}^2 \rangle}{V}\right]_v \tag{8.10}$$

Here K is a constant; V_u and V are the molar volumes of the solute and solvent, respectively; n is the refractive index of the solvent; and r_j is the electron cloud distribution of an electron in the solvent molecule, the sum being taken over all electrons of the molecule. Based on this theoretical expression, values for the solvent van der Waals term were estimated from information taken from the literature. Using target transformation, the results shown on the extreme right in Table 8.3 were obtained. Considering the crudeness of the model, the fit is surprisingly good.

8.1.7 Combination of the Basic Solvent Factors

In target factor analysis each basic vector can be tested individually; therefore, we can focus attention on any one fundamental factor. The ultimate test of the theoretical model requires a simultaneous transformation into the three fundamental solvent factors; unity, $\sigma_w(v)$, and $\sigma_a(v)$, in accord with (8.6). Simultaneous transformation into these three test vectors yielded the following equations:

$$\delta^{HMD,X}(CH_4, v) = -10.4(1) + 103.8\sigma_w(v) + 1.10\sigma_a(v)$$
$$\delta^{HMD,X}(CH_3CH_3, v) = 34.3(1) + 72.5\sigma_w(v) + 1.07\sigma_a(v)$$
$$\delta^{HMD,X}(\text{neo-}C_5H_{12}, v) = 40.2(1) + 65.0\sigma_w(v) + 1.03\sigma_a(v)$$
$$\delta^{HMD,X}(C_6H_{12}, v) = 74.7(1) + 45.4\sigma_w(v) + 1.04\sigma_a(v) \qquad (8.11)$$
$$\delta^{HMD,X}(C_8H_{16}, v) = 82.9(1) + 39.0\sigma_w(v) + 1.06\sigma_a(v)$$
$$\delta^{HMD,X}(TMS, v) = 14.2(1) + 63.5\sigma_w(v) + 1.04\sigma_w(v)$$

These equations reproduced the measured shifts within experimental error, ±1.5 Hz.

The validity of this analysis is also substantiated by the following observations concerning the loadings (i.e., the solute coefficients) in the equations above. The coefficients of the unity factor compare quite favorably with the measured gas-phase shifts. Also, the solute coefficients of solvent $\sigma_a(v)$ terms compare quite closely to unity. Finally, a plot of the predicted solute coefficients of the van der Waals terms in the equations above versus the solute parts of (8.10) exhibits a linear trend, with the points being somewhat scattered, however. This scatter is probably due to the fact the hydrogen atoms in different solute molecules occupy different positions relative to the center of the molecule. The van der Waals model used in the factor analytical studies does not take such site factors into account.

8.1.8 Summary

The ultimate objective of factor analysis is to convert a matrix of experimental points into a set of equations that reveal the true origin of the observations. These studies clearly demonstrate that such an objective can be reached. We have seen here how factor analysis can be used to unravel the important interactions which are responsible for proton NMR solvent shifts. Until the advent of TFA, it was impossible to isolate the important interaction effects so that each basic factor could be tested individually against prevailing theories.

The main accomplishments that resulted from applying TFA to proton solvent shifts are summarized below:

1. Solvent shifts, obtained from internal standard data, were predicted empirically without recourse to any particular theory.

2. Solute gas-phase shifts were isolated from solvent contributions by a mathematical route rather than by an experimental route.

3. The additivity model of Buckingham was shown to describe the nature of the effects of solvent on the chemical shifts of nonpolar solutes.

4. Within the accuracy of the data, it was shown that each interaction term could be expressed as a product function.

5. The solvent anisotropy effect was successfully isolated from the van der Waals shift.

6. Quantitative values of the solvent anisotropy were predicted for solvents, for which no theory exists.

7. The continuum dispersion model of Linder was redeveloped to better represent the nature of the van der Waals shift term and was target-tested successfully.

8. The combination of the three identified factors, (1) solute gas-phase shift, (2) solvent anisotropy, and (3) van der Waals' contribution, predicted the solvent shifts within experimental error.

These studies focused attention on simple nonpolar solutes. The total solution to the solvent shift phenomenon, encompassing every conceivable solvent and solute, still remains to be unraveled.

8.2 ¹H, ¹³C, and ²⁹Si Solvent Shifts of TMS and Cyclohexane

Bacon and Maciel[15] applied abstract factor analysis to the "intrinsic" solvent shifts of TMS and cyclohexane. The intrinsic shift is defined as the shift of the solute in some solvent relative to the shift of the solute in pure TMS, corrected for bulk-susceptibility differences. The study involved the ¹H, ¹³C, and ²⁹Si shifts of the two solutes in 38 solvents. Because of experimental difficulties in recording ¹³C signals in natural abundance, Bacon and Maciel did not employ pure solvents. Instead, their solvents contained 20 vol % TMS. In the case of the TMS nuclides, the shifts were referenced with respect to pure TMS. For the cyclohexane nuclides the solutions contained 2 vol % cyclohexane in addition to the 20 vol % TMS, and the shifts were referenced relative to cyclohexane, 2 vol % in TMS.

Principal factor analysis was performed on four sets of solvents. Set 1 consisted of cyclohexane, benzene, *o*-dichlorobenzene, and 1,2,4-trichlorobenzene; set 2 consisted of 14 solvents; set 3 consisted of 15 solvents, hexafluorobenzene being added to set 2; and set 4 involved all 38 solvents but only the four nuclides for which all data were available, thus excluding ²⁹Si.

The first two major eigenvectors reproduced the data of solvent sets 1, 2, and 3 within experimental error. For set 4, three factors were required, suggesting that an additional effect such as a dipole-induced dipole interaction between the solvent and solute is present in the larger scheme. In order to ascertain which solvent or solvents were principally responsible for the additional factor, a "suspicious" solvent was added to set 3 and a factor analysis was performed. A series of such tests revealed that hexafluorobenzene and carbon disulfide were the major contributors to the mysterious third factor. Solvent polarity was ruled

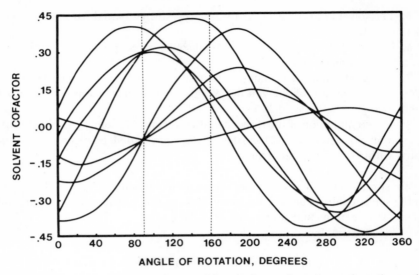

Fig. 8.1. Dependence of solvent cofactors of data set 3 upon the orthogonal rotation angle. In order of increasing value at zero degrees the solvents are iodocyclohexane, iodobenzene, bromocyclohexane, bromobenzene, chlorocyclohexane, chlorobenzene, cyclohexane, and benzene [Reprinted with permission from M. R. Bacon and G. Maciel, *J. Am. Chem. Soc.*, **95**, 2413 (1973)].

out as a cause of the third factor because acetone and chloroform, which are highly polar, had no appreciable effect on the third eigenvalue. It was postulated that perhaps the partitioning into solvent–solute pairs was beginning to break down, thus producing an additional factor.

For data sets 1, 2, and 3, a two-dimensional abstract transformation matrix, involving a single rotation angle θ, is applicable:

$$[T] = \begin{bmatrix} \cos\theta & -\sin\theta \\ \sin\theta & \cos\theta \end{bmatrix} \qquad (8.12)$$

The entire range of solvent cofactors was examined by rotating the eigenvectors at 5° increments. The results of this procedure are illustrated in Figure 8.1 for the monohalogenated benzenes and cyclohexanes. To interpret this graph in terms of two pairs of solvent–solute vectors which correspond to dispersion interaction and anisotropy, the following observations were made. An eigenvector rotation angle of 90° produces a common factor for the cyclohexanes and a different common factor for the benzenes, as illustrated by the intersecting lines in Figure 8.1. This factor was interpreted to be due to solvent anisotropy. The solute coefficients of the anisotropy factor showed a variation from one nucleotide to another: 1.04 for ^{1}H (TMS), 1.43 for ^{13}C (TMS), 1.26 for ^{1}H (CHX), and

1.11 for ^{13}C (CHX). These coefficients evidently are a quantitative measure of the various site factors of the different nuclei of the solute molecules.

A rotation angle of 160° produces a situation where the halogen substituent is important. At this angle the solvent cofactor associated with a given halobenzene is identical with the solvent cofactor associated with the corresponding halocyclohexane (see Figure 8.1). This situation, according to Bacon and Maciel, is consistent with their dispersion interaction guidelines but inconsistent with all other dispersion models.

8.3   FLUORINE SOLVENT SHIFTS

Abraham et al.[16] measured the ^{19}F shifts of 19 rigid, nonpolar solutes in eight solvents. They suggested that two factors, the solute gas-phase shift and the van der Waals effect, were responsible for the solvent shifts. When these data were subjected to factor analysis, Malinowski[17] found evidence, based upon the IE and IND functions (see Section 4.3.2), that not two but three factors were active. Although small, the third factor was suspected to be due to solvent anisotropy even though benzene and carbon disulfide, which are known to exhibit large anisotropic effects, were not among the solvents employed. There was no way of testing this effect because the anisotropies of the solvents involved were not known.

In another study, using the same data, Malinowski[18] confirmed the solute gas-phase shift to be a basic factor by target-testing the gas-phase shifts. The test vector and the predicted vector are given in Table 4.7. The theoretical target errors, also listed in the table, give evidence that the gas-phase shift is a basic factor. This study also shows that the predicted gas-phase shifts are more accurate than the measured gas-phase shifts (REP = 0.05 ppm, whereas RET = 0.19 ppm). This occurs because the solution shifts are more accurate than the gas shifts.

8.4   HALOGEN SUBSTITUENT EFFECTS ON ^{13}C SHIFTS

In order to probe into the origins of ^{13}C shifts, Wiberg et al.[19] focused attention on halogen-substituted hydrocarbons. They factor-analyzed the "differential" chemical shift (i.e., the difference in shift between the halogen-substituted compound and the unsubstituted parent compound). Their data matrix consisted of 62 differential shifts as column designees and four halogens (F, Cl, Br, I) as row designees.

Two major factors emerged from the analysis. Target tests yielded three successful halogen property vectors: a unity vector (1   1   1   1), an arithmetic progression vector (1   2   3   4), and a bimodal vector (1   0   0   1). Upon using these three vectors in combination, three associated loading cofactors were obtained for each carbon nuclei involved. The three different types of loadings were believed to be due, respectively, to a "hypothetical" halogen having the same quantum number, the "freeness" of the valence electrons, and a "conformation" effect.

REFERENCES

1. J. Homer, *Appl. Spectrosc. Rev.,* **9,** 132 (1975).
2. P. H. Weiner, E. R. Malinowski, and A. R. Levinstone, *J. Phys. Chem.,* **74,** 4537 (1970).
3. P. H. Weiner and E. R. Malinowski, *J. Phys. Chem.,* **75,** 1207 (1971).
4. P. H. Weiner and E. R. Malinowski, *J. Phys. Chem.,* **75,** 3160 (1971).
5. P. H. Weiner, Ph.D. thesis, Stevens Institute of Technology, Hoboken, N.J.; *Diss. Abstr.,* **32** (6), Publ. No. 3290B (1971).
6. A. D. Buckingham, T. Schaefer, and W. G. Schneider, *J. Chem. Phys.,* **32,** 1227 (1960).
7. E. R. Malinowski and A. R. Pierpaoli, *J. Magn. Reson.,* **1,** 509 (1969).
8. A. A. Bothner-By, *J. Mol. Spectrosc.,* **5,** 52 (1960).
9. W. T. Raynes, A. D. Buckingham, and H. J. Bernstein, *J. Chem. Phys.,* **36,** 3481 (1962).
10. J. C. Schug, *J. Phys. Chem.,* **70,** 1816 (1970).
11. J. Homer, *Tetrahedron,* **23,** 4065 (1967).
12. B. Linder, *J. Chem. Phys.,* **33,** 668 (1960).
13. B. B. Howard, B. Linder, and M. T. Emerson, *J. Chem. Phys.,* **36,** 485 (1962).
14. F. H. A. Rummens, W. T. Raynes, and H. J. Bernstein, *J. Phys. Chem.,* **72,** 2111 (1968).
15. M. R. Bacon and G. Maciel, *J. Am. Chem. Soc.,* **95,** 2413 (1973).
16. R. J. Abraham, D. F. Wileman, and G. R. Bedford, *J. Chem. Soc. Perkin Trans. II,* **1973,** 1027.
17. E. R. Malinowski, *Anal. Chem.,* **49,** 612 (1977).
18. E. R. Malinowski, *Anal. Chim. Acta,* **103,** 339 (1978).
19. K. B. Wiberg, W. E. Pratt, and W. F. Bailey, *Tetrahedron Let.,* **1978** (49), 4861, 4865.

9 Chromatography

9.1 INTRODUCTION

In this chapter we show how factor analysis has been used to broaden our understanding of chromatographic processes. In particular, factor analysis has been employed to identify basic factors which influence solute–solvent interactions and to classify solutes and solvents into groups having similar characteristics. Results from nearly 20 factor analyses of chromatographic data are discussed.

Factor analytical solutions to chromatographic retention problems have the general form

$$d_{ik} = \sum_j u_{ij}v_{jk} \qquad (9.1)$$

where d_{ik} is the measured retention for the ith solute chromatographed on the kth stationary-phase solvent, and u_{ij} and v_{jk} are the jth cofactors for the solute and solvent, respectively. In areas such as gas–liquid chromatography (GLC) and liquid–liquid chromatography, the stationary phase is an immobile liquid which acts as a solvent for solutes in the mobile phase. The matrix expression corresponding to (9.1) is

$$[D] = [U][V] \qquad \cdot (9.2)$$

where $[D]$ is the matrix of chromatographic retention data, and $[U]$ and $[V]$ are the solute- and solvent-cofactor matrices, respectively.

The measured retention represents a composite of all interactions that govern the movement of a solute through the chromatographic column. The energetics of solute–solvent interactions in chromatography are described in Section 2.2 of Karger et al.'s monograph.[1] The ubiquitous interaction in chromatography is due to London dispersion forces arising from induced dipole-induced dipole interactions. Dispersion interactions play the dominant role in selective retardation of nonpolar solutes. Polar interactions involving dipole-induced dipole forces and dipole–dipole forces become major factors if either the solute or the solvent are polar. In some separations, interactions involving electron donor-acceptor forces such as hydrogen bonding play a crucial role. In special circumstances, steric factors can assume significance.

Nearly all of the applications of factor analysis to chromatography have involved GLC data. Two methods for reporting GLC data, based on the specific retention volume and the retention index system, are appropriate for factor analysis. The logarithm of the corrected specific retention volume, V_r^0, is proportional to the standard molar free energy of solution for chromatographic distribution phenomenon. Since careful control is required to determine accurate values of V_r^0, the retention index system has become the preferred method for presenting GLC results. Retention volumes of solutes in an homologous series are used as reference points in the retention index approach. A solute's position on this scale is measured relative to those of the marker solutes. The retention index I_{ik} of solute i on stationary-phase solvent k is defined as

$$I_{ik} = 100n + 100 \frac{\log V_i - \log V_n}{\log V_{n+1} - \log V_n} \qquad (9.3)$$

Here, n and $n + 1$ are the numbers of carbon atoms in the marker solutes eluted directly before and directly after solute i, and V_i, V_n, and V_{n+1} are the retention volumes of the ith solute and the two marker solutes, respectively. Based on thermodynamic arguments of Rohrschneider,[2] we expect that the retention index can be expressed as a linear sum of products and therefore that matrices of retention indices will have FA solutions.

9.2 SUMMARY OF PROBLEMS

In this section we present an overview of the applications of factor analysis to chromatography. A total of 22 problems, involving 19 publications, will be summarized.

Scope. Factor analytical contributions in chromatography have been of two main types. The majority of the applications utilize target factor analysis to acquire knowledge about the factors that govern GLC separations. In a second type of application, both target factor analysis and abstract factor analysis have been employed to classify solutes and stationary-phase solvents into groups having similar behavior.

A concise summary of the factor analysis/chromatography studies is presented in Table 9.1. Information concerning the nature of the data matrices is given in the first three columns of table. With the exception of two problems based on liquid–liquid chromatographic data,[10,13] the problems utilized GLC data. The extensive compilations of McReynolds[23,24] have been especially useful sources of GLC data. The retention index, with but one exception,[3] has been the type of GLC data factor-analyzed. The number of rows and columns in the data matrices has generally been much greater than the number of factors (see Section 9.4). A variety of organic solutes and solvents have been incorporated in the studies. Some problems[6-8,10,12,13,15,21] were restricted to a single class of solutes; the remaining problems involved solutes from several different families. Stationary-phase solvents in most cases spanned a wide range of polarities. In all but four problems,[11,19-21] the solvents were primarily polymeric. The type of factor analysis employed is given in the fourth column of Table 9.1. Fifteen problems involved target factor analysis; the other problems involved abstract factor analysis. The last column in the table lists the references of each paper. The factor analytical applications were due mainly to two researchers, Howery and Weiner. In a series of pioneering collaborations,[4-6,14] Howery and Weiner demonstrated that target factor analysis furnished insights into the basic factors affecting GLC retention.

Information. A summary of information from GLC applications utilizing abstract factor analysis only is given in Table 9.2. These papers will be discussed in Section 9.3.

Information from target factor analyses is summarized in Table 9.3. These papers will be discussed in Section 9.4 and following. Ranks of the data matrices,

Table 9.1 Summary of applications of factor analysis to chromatography

		Data Matrix		
Data[a]	Solutes[b]	Solvents	Type of FA[c]	Reference
log a	39	7 polymeric	A	3
I	30	23 polymeric	T	4
I	30	23 polymeric	T	5
I	26 alcohols	25 polymeric	T	6
I	15 esters	6 polymeric	T	7
I	15 esters	6 polymeric	T	8
I	26 alcohols	25 polymeric	T	8
I	10	210 polymeric	A	9
log K	26 steroids	6 polymeric	A	10
I	10	5 mixed	T	11
I	25 hydrocarbons	12 polymeric	T	12
log V	15 carboranes	8 polymeric	T	13
I	44	25 polymeric	T	14
I	39	12 polymeric	T	14
I	18 ethers	25 polymeric	T	15
I	10	225 polymeric	A	16
I	10	226 polymeric	A	17
I	30	23 polymeric	A	18
I	10	226 polymeric	A	18
I	49	7 monomeric	T	19
I	53	18 monomeric	T	20
I	39 carbonyls	18 monomeric	T	21

[a] a, activity coefficient; I, retention index; K, liquid-chromatographic partition coefficient; V, liquid-chromatographic retention volume.
[b] Incorporates a variety of organic solutes unless otherwise indicated.
[c] A, abstract factor analysis; T, target factor analysis.

Table 9.2 Summary of abstract factor analyses

Reference	Factors	Main Objective
3	5	Prediction of new data
9	3	Classification of solvents
10	3	Calculate abstract factors
16	2, 3	Classification of solvents
17	2	Classification of solvents
18	2–5	Classification of solutes and solvents

190

Table 9.3   Summary of information from target factor analyses

Reference	Factors[a]	Uniqueness Tests[b]	Typical Designees[b,c]	Target Transfor- mations[b]	Key Basic Factors[b,d]	Predictions[e]
4	8	—	—	u	—	—
5	8	u, v	u, v	u, v	—	t
6	5, 6[f]	u, v	—	u	u	b
7	4	—	v	—	—	—
8	4	u	—	u	u	e
8	5	—	v	—	—	e
11	1, 2[f]	—	—	v	v	—
12	6	u	u	u	u	—
13	3	—	—	u, v	u	e
14	1–8[f]	u	u	u	—	—
14	1–3[f]	—	—	—	—	—
15	6	u, v	u, v	u	u	bg
19	3	u, v	u, v	u, v	u, v	b
20	4–7[f]	u, v	u, v	u, v	u, v	—
21	4–7[f]	u, v	u, v	u, v	u, v	—

[a] Based on number of abstract factors to reproduce data near experimental error.
[b] u and v denotes results published for solutes and solvents, respectively; a blank signifies that the step was not carried out.
[c] Based on combination step using typical vectors.
[d] Based on combination step using basic vectors.
[e] t, based upon solvent-associated typical vectors; b, based upon basic vectors; e, based upon matrix expanded with a basic vector (see Section 9.7).
[f] Depends upon submatrix studied (see original articles).
[g] Results given in reference 22.

estimated from the reproduction step, are shown in the second column of Table 9.3 (see Section 9.4 for discussion). The next five columns in the table indicate which of the specified steps of TFA were carried out in each problem. The symbols u and v signify that the designated TFA procedure was performed for solutes and solvents, respectively. For example, for the third problem[6] in the table, uniqueness values for both solutes and solvents were determined; key, typical vectors were not determined for either solutes or solvents; and target tests for basic vectors, combinations of basic vectors, and predictions based on basic vectors were published for solutes only.

General conclusions and representative examples from each step of factor analysis will be discussed in the following sections. The order follows nearly the order of presentation in Chapter 6. To keep the discourse in perspective, the reader will find Tables 9.1, 9.2, and 9.3 useful.

9.3 ABSTRACT FACTOR ANALYSES

Applications of abstract factor analysis to chromatography are listed in Table 9.2. Funke et al.[3] were the first to apply factor analysis to chromatography. They used the principal factor solution to predict activity coefficients. Key sets of five typical solutes and of five typical solvents were chosen from chemical criteria. The cofactor coefficients needed for the predictions were taken from the principal factor model rather than from a combination-TFA solution. Activity coefficients for a new solute and for five new solvents were predicted quite favorably. Although target testing was not employed, the paper foretold the essential features of target factor analysis.

The main goal in the remainder of the abstract factor analyses listed in Table 9.2 was to classify solvents. There exists a profusion of stationary-phase solvents. Hence mathematical methods for classifying solvents and for detecting redundancies among GLC packings are of practical interest. The ultimate goal is to identify a small set of key, preferred solvents which can accomplish the vast majority of GLC separations.

Abstract factor analysis has been used to identify major similarities and differences in stationary phases. Clusterings were based upon the similarities of cofactors in the principal factor solution. Unfortunately, rotational techniques such as varimax (see Section 6.4) have not been applied to GLC data. Users of AFA have employed the percent variance (see Section 4.3.2) to estimate the number of factors. Typically, an AFA–GLC solution has been considered adequate if say 98% of the variance is accounted for. Two-factor or three-factor models have been used for classification purposes. (For target factor analysis, the variance can be deceiving since, even if 99% of the variance is accounted for, GLC data often are not reproduced near experimental error.)

In the ground-breaking use of AFA to categorize solvents, Wold and Andersson[9] selected a three-factor model as a reasonable compromise between generality and practicality. Three principal components reproduced the data within about 30 r.i. units. The primary abstract factor was attributed to the polarity of the solvent, the second factor was accounted for by an unspecified but relatively constant solute parameter, and the third factor was attributed to hydrogen-bonding interactions involving alcoholic solutes.

The approaches employed by McCloskey and Hawkes,[16] Lowry et al.,[17] and Chastrette[18] were similar. In order to describe the solvents in an easily visualized manner, two-factor solutions were utilized. Plots of solvent cofactor 1 against solvent cofactor 2 indicated clusterings of solvents. Each point in such a plot represents a single solvent. The utility of this approach is illustrated in Figure 9.1. The clusterings of 225 stationary-phase solvents depicted in the figure are consistent with chemical insight. Preferred stationary phases are likely to be those which have cofactors far apart on both axes of the two-dimensional plots.[17]

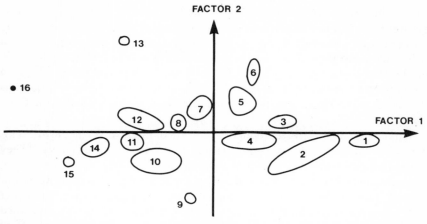

Fig. 9.1. Two-dimensional plot of principal cofactors for 225 solvents. Types of solvents in each cluster: 1, apolars; 2, silicons; 3, alkyl adipates; 4, phthalates; 5, Ucon-related; 6, amides; 7, Ethofat-related; 8, Igepal-related; 9, fluorinated; 10, cyano-containing; 11, esters of polyols; 12, Carbowax-related; 13, Siponat and Stepan; 14, succinates; 15, cyanoethoxypropane-related; and 16, diglycerol [Reprinted with permission from M. Chasterette, *J. Chromatogr. Sci.,* **14,** 357 (1976)].

Redundant solvents form clusters in the two-factor space. In another use of the two-dimensional plots, Chastrette[18] suggested that the plot for solute cofactors be superposed on the plot for solvent cofactors. Solute–solvent pairs which interact the most with each other tended to have points near each other on the superposition.

9.4 NUMBER OF FACTORS

The number of factors found for 15 different target factor analyses varied between one and eight, as shown in Table 9.3. As the complexity of the problems increased, the number of factors generally increased. Only one or two factors were required for problems that were considered to be chemically simple.[11,14] In the most complicated problems,[4,14,20,21] as many as seven or eight factors were needed. In some cases[15,19-21] the ranks were substantiated using the theory of error for abstract FA discussed in Chapter 4.

The average error or the RMS error of data reproduction was usually employed to estimate the number of factors. For example, Selzer and Howery[15] reported that, with five factors, the average error of reproduction was 2.8 r.i. units, the largest error was 14 r.i. units, and about 8% of the data were reproduced with errors exceeding 5 r.i. units. With six factors, the average error dropped to 2.1, the largest error decreased to 8.2, and only 3.6% of the data had errors greater than 5. Since the experimental error had an upper limit of 5 r.i. units, six factors were deemed necessary. Malinowski[25] reached the same conclusion using the factor indicator function (IND; see Section 4.3), which exhibited a mininum at six factors. In other GLC problems, Howery and Soroka[19-21] also found that the number of factors based on the experimental-error criterion was the same as the number of factors deduced from the factor indicator function.

Three other uses of the reproduction step are of interest. Weiner et al.[11] employed the dimensions of the factor space to verify retention mechanisms. They found that one factor was required if only nonpolar solutes were considered, whereas two factors were needed if both polar and nonpolar solutes were considered. The number of factors estimated from factor analysis was consistent with a chemical model for the retention mechanism. Howery and coworkers used the reproduction step to furnish insights into the effects of solutes on the complexity of data spaces. They studied the change in data reproduction when columns of data were added to or deleted from the data matrix. In one approach, Howery et al.,[14] starting with a data matrix containing only alkane solutes, determined the increase in the factor size when data columns for one or two new solutes were added to the data matrix. The increases in rank were related to the complexities of the functional groups of the new solutes. For example, a solute with an alkyl side chain had almost no effect on the factor size, while an alcohol solute increased the rank by nearly 1 unit. Using the opposite approach, Howery and Soroka[19] studied the effects of removing one or more solutes from the data matrix. Usually, the greatest reduction in rank was observed for polar solutes and for chemically unique solutes.

9.5 UNIQUENESS AND UNITY TESTS

The uniqueness test (see Section 6.7.2) is a useful, preliminary target test which enables the chromatographer to ferit out designees which are responsible for special interactions or which contain gross errors. Uniqueness tests can also be employed to identify clusters between similar solvents.

A summary of solutes and solvents that yielded high uniqueness values is tabulated in Table 9.4. In most cases the unique molecules have either polar functional groups or small molecular masses. For example, the pronounced uniqueness of the solvent diglycerol[6,15,20,21] is indicative of the hydrogen-bonding capability of diglycerol. High uniqueness values, not anticipated from chemical insight, might indicate gross errors. With this observation in mind, Weiner and Parcher[8] employed uniqueness tests to single out suspicious data points. Two esters were found to be quite unique (see Table 9.4), yet chemically the esters

Table 9.4   Summary of unique solutes and solvents

	Reference
Solutes[a]	
Pyridine (0.73), *t*-butanol (0.56)	5
2-Propynl-1-ol (0.81), methanol (0.71)	6
Isobutylbutyrate[b] (1.00), isopropylisobutyrate[b] (0.81)	8
Ethynlbenzene (0.90), 2-octyne (0.74), cyclohexane (0.70)	12
2-Propynl-1-ol (0.75), octadecane (0.55)	14
Dimethyl ether (0.84), isobutylmethyl ether[b] (~0.8), *t*-butylmethyl ether (0.54)	15
Acetaldehyde (0.51)	21
Solvents[a]	
Fluorosilicone QF-1 (0.85), diethyleneglycol succinate (0.60), silicone fluid nitrile XF 1150 (0.60)	5
Diglycerol (0.93), polyphenyl ether—5 rings (0.56)	6
Diglycerol (0.90), Hyprose (0.60), Zonyl E7 (0.60)	15
Hexadecyl iodide (0.90), heptadecane (0.65)	19
Diglycerol (0.99), Zonyl E7 (0.94), sucroseoctaacetate (0.61), Quadrol (0.55)	20
Diglycerol (0.99), Zonyl E7 (0.98), sucroseoctaacetate (0.83), Hyprose (0.66)	21

[a] Predicted uniqueness value given in parentheses.

[b] Designee not included in final TFA (see original paper for discussion).

were not considered to be atypical. Suspected of having gross experimental error, these two esters were eliminated from the data matrix.

A concise table can be employed to summarize the results of uniqueness tests. Weiner and Howery[5] published such a table for a number of solute uniqueness tests. Representative results are shown in Table 9.5. Pyridine and *t*-butanol, both relatively unique chemically, had the largest uniqueness values. Solutes with fairly high predicted values on a given uniqueness test were assumed to form a cluster of similar solutes. For example, although benzene, toluene, and styrene were not unique individually, these chemically similar solutes were clustered as indicated in the table. Uniqueness tests do not necessarily cluster designees which are chemically alike, as shown by the alcohols in Table 9.5.

In GLC studies the results of unity tests (see Section 6.7.3) have been consistent with chemical expectations. When all the solutes contain at least one real factor in common, such as the same functional group, unity should be a factor. For problems involving solutes from one family,[6,8,13,16,21] unity transformed

Table 9.5   Uniqueness table for solutes [a]

Solutes Tested	Uniqueness Value	Other Solutes Clustering with Tested Solute
2,4-Dimethyl pentane	0.34	Cyclohexane
Cyclohexane	0.25	2,4-Dimethyl pentane
Benzene	0.16	Toluene, styrene
Toluene	0.18	Benzene, styrene
Styrene	0.26	Toluene
Methyl iodide	0.27	Cyclohexane
Ethyl bromide	0.07	Methyl iodide
Acetone	0.17	Methyl ethyl ketone
Methyl ethyl ketone	0.19	Acetone
Propionaldehyde	0.11	Acetone, methyl ethyl ketone
Nitromethane	0.35	Nitroethane
Nitroethane	0.23	Nitromethane
Pyridine	0.73	—
Ethanol	0.19	—
2-Propanol	0.18	Ethanol, *t*-butanol
t-Butanol	0.56	2-Propanol

[a] Reprinted with permission from P. H. Weiner and D. G. Howery, *Anal. Chem.,* **44,** 1189 (1972).

quite well. In most of the other studies referred to in Table 9.2, the unity vector was not a factor.

9.6   KEY SETS OF TYPICAL VECTORS

To search for a key set of typical vectors, columns of the data matrix are employed in combination TFA (see Section 6.5). For GLC, a summary of the best key sets is presented in Table 9.6. In two problems,[7,15] every possible combination of typical vectors was tested.

The results demonstrate that retention index data can be represented adequately by small sets of typical vectors. As shown in the right-hand column of Table 9.6, every acceptable key set reproduced the data matrix near or within experimental error. Molecules represented in the key sets were usually similar to those which might be selected on the basis of chemical insight.[7] Many combinations gave almost equivalent results, indicating considerable redundancy of both solutes and solvents. Testing all combinations has the advantage of identifying the best solution. Selzer and Howery[15] designed a pattern table which

Table 9.6 Summary of key combinations of typical vectors

Reference	Scope[b]	Error, TFA solution[a]		Experimental Error[c]
		Key Set Solutes	Key Set Solvents	
5	s	4	3	5
7	a	—	2.5	2
8	s	—	2.5	5
12	s	0.8	—	1
14	s	2.1	—	5
15	a	2.8	2.9	5
19	a	0.7	0.6	1
20	s	2.4	2.0	5
21	s	2.3	2.2	5

[a] Average or RMS error (r.i. units) for key combination.
[b] s, selected combinations; a, all combinations.
[c] Estimated upper limit of experimental error (r.i. units).

listed the precentage of times each typical vector occurred in the better combinations for a range of factors. By examining the trends in the pattern table, they were able to estimate the relative importance of each vector.

9.7 PREDICTIONS BASED ON KEY SETS OF TYPICAL VECTORS

Predictions of new retention data based on key combinations of typical vectors (see Section 6.6) have been accomplished in three GLC studies.[5,8,13]

Weiner and Howery[5] used typical vectors to predict retention indices for four new solutes on 15 nonkey solvents. Retention data for each of the new solutes on eight key solvents were required because the factor space was found to be eight-dimensional. Retention indices were predicted with an average error of 4.4 r.i. units, within the upper limit of the experimental error.

Weiner and Parcher[8] introduced a novel procedure from which new values of basic cofactors, rather than new retention indices, were predicted. They expanded the data matrix by adding a complete basic vector to the data matrix as the $(c + 1)$th data column. The key set of typical solvent vectors for the original data matrix was determined by combination in the usual manner. The enlarged matrix was then combination-reproduced with the same set of key typical vectors to obtain a target-transformed column matrix. Then data were predicted for new row-designee solvents. For nonkey columns up to the cth column, new retention data resulted. More intriguingly, for the $(c + 1)$th column, values of the basic parameter corresponding to the added column were predicted for the new solutes. Predictions based on the expanded-matrix method were considered valid only if the dimensionalities of the original data matrix and the expanded matrix were identical. The method was used by Weiner and Parcher[8] to predict values for 10 chemical parameters, such as boiling points, carbon numbers, and van der Waals' constants. Agreement between predicted and known values was quite satisfying. Kindsvater et al.[13] used the unity vector as the added column and found that the predicted values were close to unity only if the new solute was chemically similar to the original solutes.

9.8 TARGET TESTS

Testing parameters of solutes and solvents via target transformation has been the most fundamental aspect of the factor analytical investigations of chromatographic data.

9.8.1 Overview of Successful Tests

Properties of solutes and, to a lesser extent, of solvents have been target tested in several GLC studies, as shown in Table 9.3. An overview of possible basic factors based on successful target tests is presented in Table 9.7. Parameters that occur more than once in the table are abbreviated in the tabulation and identified in a footnote.

In the pioneering application of TFA to gas chromatography, Weiner and Howery[4,5] studied a complex data set which contained a diversity of solute and solvent types. They found that target testing isolated many chemically reasonable factors. As GLC–TFA researchers have gained experience in developing test vectors, the scope of target testing has increased. For example, Selzer and Howery[15] tested over 50 solute vectors, and in addition tested the square, reciprocal, and logarithm of each vector.

Several conclusions can be drawn from an examination of the solute factors listed in Table 9.7. Significantly, a large number of test parameters have been found to be factors in more than one problem. Three real vectors—carbon number (CN), molar refraction (MR), and molecular weight (MW)—were judged to be solute factors in at least eight problems. Chemically, each of these factors might be related to dispersion interactions. Other factors being constant, elution according to molecular weight is a well known empirical rule of GLC.

Table 9.7   Summary of basic factors identified by target testing

	Reference
A. Basic Solute Factors[a]	
HU,[b] nitrogen uniqueness,[b] VW[b]	2
DM, square DM,[b] EV, MR,[b] VW	3
BP,[b] CN,[b] DC, DM,[b] EV,[b] hydroxyl position, log VP, MR,[b] MW,[b] UN[b]	4
BP, CN, EV,[b] log VP,[b] MR,[b] MW, UN,[b] VW[b]	6
AU,[b] carbon-bond scale,[b] CN,[b] enthalpy of formation, EV, heat capacity,[b] heat content, HC,[b] MR, NV,[b] MW, multiple-bond scale,[b] multiple-bond uniqueness,[b] triple-bond uniqueness[b]	11
CN,[b] DM, square DM, log GLC, retention volume[b]	12
CN,[b] DE, DM, EV, MR,[b] MW, log VP, VW[b]	13
AN, BP, square BP,[b] CN, chain-related vectors, CN difference, freezing point, methyl uniqueness,[b] methyl ether uniqueness,[b] MR, MW, UN, unsaturation uniqueness,[b] vinyl uniqueness	15[c]
AN, BP, square BP,[b] CN, DE, DM, EV, GD,[b] IU, MR, MW, RD,[b] UN, XU	20
AU, BP, square BP, CN, cyclic uniqueness, DC, EV, HC, HU, ketone uniqueness, MS,[b] melting point, MR, MV,[b] MW, ON, UN	21
AU, BP, square BP, CN, DM, group DM, EV, HC, HU, MS, MW, surface tension, VW	22
B. Basic Solvent Factors[a]	
Ether-like uniqueness[b]	2
Interfacial area,[b] bulk volume[b]	10
DC,[b] UN[b]	12
DM,[b] GD,[b] halogen-ether uniqueness, IU, MR, MW, RD,[b] UN	20[c]
AN, EU, HU, MC, MR, MW, SU, UN	21[c]
AN, AU, EU, HU, MC,[b] MR,[b] ON,[b] SU, UN, XU	22[c]

[a] Vectors appearing more than once in the table are abbreviated as follows: AN—atom number; AU—aromatic uniqueness; BP—boiling point; CN—carbon number; DC—dielectric constant; DE—density; DM—dipole moment; EU—ether uniqueness; EV—enthalpy of vaporization; GD—group delta; HC—heat of combustion; HU—hydroxyl uniqueness; IU—iodine uniqueness; MC—McReynolds constant; MR—molar refraction; MS—magnetic susceptibility; MV—molecular volume; MW—molecular weight; ON—oxygen number; RD—r.i. dispersion; UN—unity; VP—vapor pressure; VW—van der Waals' constant; XU—halogen uniqueness.

[b] Details of target test are given in the reference.

[c] Additional vectors are discussed in the reference.

Parameters related to the vaporization of the solutes have frequently tested well. Rohrschneider[2] postulated that the retention index is proportional to the free-energy change for solute retention and therefore to the enthalpy of solution. As pointed out by Weiner and Parcher,[8] the enthalpy of solution can be divided into the heat of vaporization for the solute and the heat of mixing. From Table 9.7 we see that the enthalpy of vaporization (EV) is a solute factor in eight problems. On the basis of chemical thermodynamics, parameters proportional to the enthalpy of vaporization, such as the boiling point ($°K$) and the logarithm of the vapor pressure, should be factors. Satisfyingly, the boiling point has been reported as a solute factor in six studies, and log VP was identified in three studies.

Another type of factor accounts for polar interactions. Dipole moment (DM) tested adequately in five problems and the square of the dipole moment was a basic factor in three problems. Both forms of the dipole moment appear in thermodynamic interaction terms.[1]

Other types of factors have been suggested from TFA results. Weiner and Howery[4] showed that gas-phase nonideality can be related to the van der Waals constants. These constants tested well in five GLC problems. That target factor analysis could isolate such a minor factor attests to the sensitivity of target testing. The unity vector, which represents a constant solute property, passed the target test in five problems involving solutes with a common functional group.

Most of the other factors in Table 9.7 are specific kinds of structural parameters. Clustering uniqueness vectors for solutes having similar chemical properties usually tested successfully. For instance, uniqueness tests for aromatic solutes indicated that the benzene ring was responsible for a factor in three problems. The square of the boiling point tested very well and seemed to be a major factor in several problems, in agreement with Bach et al.[26]

Parameters that chromatographers would not expect to be factors, such as melting point, refractive index, surface tension, and viscosity, generally did not transform satisfactorily. Other examples of unsuccessful tests can be found in the papers referenced in Table 9.3.

In summary, applications of TFA to retention indices indicate an intuitively acceptable model for the solute part of the solute–solvent interaction space. Target FA has been used to identify solute factors related to dispersion interactions, enthalpy of vaporization, polar interactions, gas-phase imperfections, and specific structural parameters. Overall, the high correlation between TFA results and chemical intuition is quite encouraging.

Only a few detailed studies of solvent factors have been conducted. Basic solvent parameters identified by Howery and Soroka[20-22] are shown in the lower portion of Table 9.7. The parameters listed are quite similar to the kinds of basic solute parameters discussed above. Additional target factor analytical studies of solvent factors are needed to develop better models for solute–solvent interactions in gas chromatography.

9.8.2 Solute Factors

In this section we examine results of 10 representative target tests of solute parameters. The results, each involving 13 typical solutes, are shown in Table 9.8. The solutes in these tests were selected to span a wide range of chemical characteristics.

Examples A through D in Table 9.8, taken from four different investigations, show results of testing four physical parameters: molar refraction, molecular weight, van der Waals' b constant, and dipole moment squared. Examples E through J illustrate the versatility of structural test vectors. The table shows results concerning a uniqueness test for alcohols, a uniqueness test for unsaturation, a carbon-number test, a test for a multiple-bond scale, a test for hydroxyl position, and a unity test. The advantages of free-floating test points are illustrated in examples A, E, F, and I. In each example the validity of the test was substantiated because known values of points which were deliberately free-floated were predicted quite well.

Most of the examples in Table 9.8 are self-explanatory. Examples E and I have special historical significance. The uniqueness test for alcohols in example E was the first use of target testing to single out a subgroup of designees. In this test Weiner and Howery[4] assigned values of 1 to two of the alcohols, free-floated the test values for most of the remaining solutes, and assigned a 0 to each of the hydrocarbons. The predicted values for all of the alcohols were near unity except for t-butanol. The somewhat lower predicted value for t-butanol and the rather high predicted value for pyridine were attributed to the high uniqueness of the two solutes (see Table 9.3). Example I shows an early attempt to test detailed structural parameters with TFA. In order to learn whether the position of the hydroxyl group in a series of alcohols was a factor, Weiner et al.[6] constructed a structural vector that indicated the number of carbon atoms attached to the carbon atom bonded to the hydroxyl group. The vector was essentially a test for primary, secondary, and tertiary character, with the proviso that methanol should have a value of zero on the vector. Free-floating methanol and all of the secondary alcohols was a severe test of the idea. As can be seen from the results for example I, the vector tested quite well and the free-floated points were predicted within chemical expectations. The negative values predicted for the free-floated unsaturated alcohols were believed[6] to be caused by the inductive effects of the multiple bond on the hydroxyl group.

Chemically meaningful scales have been developed using the free-floating capability of TFA. For example, Klingen et al.[13] (see example G of Table 9.8) assigned carbon numbers to five straight-chain solutes and free-floated all other

Table 9.8 Examples of target tests of solute vectors[j]

Solute	Test	Predicted
A. Molar refraction[a]		
Hexane	29.9	28.8
Octadecane	86.2	86.5
Benzene	26.2	26.3
Ethyl benzene	(35.8)	34.5
Ethyl ether	22.5	23.0
Propylmethyl ether	22.0	23.0
Methyl acetate	(17.5)	18.7
Ethyl propionate	26.8	26.4
Acetone	16.2	15.8
2-Hexanone	30.0	30.1
Propanol	17.5	17.7
Octanol	40.7	40.8
Cyclohexanol	(28.7)	29.6
B. Molecular weight[b]		
Acetaldehyde	44	41
Butanal	72	74
Isopentanal	86	84
2,2-Dimethyl propanal	86	89
Hexanal	101	101
2-Ethyl hexanal	128	128
Acetone	58	59
3-Pentanone	86	87
3,3-Dimethyl-2-butanone	100	101
2-Octanone	128	127
Cyclopentanone	84	85
2,3-Butadione	86	87
2,4-Pentadione	100	94
C. van der Waals' b constant[c]		
Methyl butyrate	0.166	0.167
Ethyl butyrate	0.194	0.194
Propyl butyrate	0.223	0.223
Butyl butyrate	—	0.250
Pentyl butyrate	—	0.279
Isopropyl butyrate	—	0.213
Isopentyl butyrate	0.274	0.274
Ethyl isobutyrate	0.189	0.186
Propyl isobutyrate	0.216	0.220
Butyl isobutyrate	—	0.246
Pentyl isobutyrate	—	0.273

Table 9.8   Continued

	Test	Predicted
Isobutyl isobutyrate	0.240	0.237
Isopentyl isobutyrate	—	0.269
D. Dipole moment squared[d]		
Cyclohexane	0.0	0.7
Toluene	0.1	0.1
Acetone	8.4	8.9
Crotonaldehyde	13.5	10.8
Butyl acetate	3.4	5.0
Acetonitrile	14.8	13.9
Nitroethane	12.6	13.0
Dibutyl ether	1.5	−0.3
Chloroform	1.0	1.6
Methyl iodide	2.6	2.8
Ethyl bromide	4.3	3.6
2-Propanol	2.6	2.3
Pyridine	4.6	4.4
E. Hydroxyl uniqueness[e]		
2,4-Dimethyl pentane	0	0.00
2-Ethyl hexene-1	0	0.00
Toluene	0	0.00
Acetone	—	−0.13
Dibutyl ether	—	0.03
Chloroform	—	0.37
Pyridine	—	0.69
Ethanol	1	1.00
Propanol	1	1.00
Isopropanol	(1)	0.94
t-Butanol	(1)	0.74
Cyclopentanol	(1)	0.95
Allyl alcohol	(1)	1.02
F. Unsaturation uniqueness[f]		
Dimethyl ether	0	0.01
Propylmethyl ether	0	0.04
Butylmethyl ether	0	0.19
t-Butylmethyl ether	0	−0.05
Butylethyl ether	(0)	0.04
Dipropyl ether	(0)	−0.03
Isopropylpropyl ether	0	−0.04
Diphenyl ether	0	0.07

Table 9.8   Continued

	Test	Predicted
Diisopentyl ether	0	−0.06
Ethylvinyl ether	1	1.07
Isobutylvinyl ether	1	1.03
2-Ethyl-1-hexylvinyl ether	1	1.01
Allylethyl ether	1	0.74
G. Carbon number[g]		
o-Carborane (θ)	0	0.07
$H\theta CH_3$	1	1.03
$H\theta C_2H_5$	2	1.90
$H\theta C_4H_9$	4	3.86
$H\theta C_6H_{13}$	6	6.09
$H\theta CHCH_2$	—	1.50
$H\theta C_2H_4CHCH_2$	—	3.03
$H\theta CH_3CCH_2$	—	2.27
$CH_3\theta CH_3$	—	1.83
$H\theta CH_2Br$	—	1.65
$CH_3\theta CH_2Br$	—	2.29
$H\theta C_6H_5$	—	3.10
$H\theta CH_2C_6H_5$	—	3.06
H. Multiple bond scale[h]		
Ethane	0	0.00
Hexane	0	0.02
2-Methane heptane	0	0.02
Decane	0	0.04
1-Octene	1	0.96
2-Ethyl hexene	1	0.95
1-Octyne	—	3.31
2-Octyne	—	3.01
Benzene	3	3.02
Toluene	3	2.98
Mesitylene	3	2.96
Styrene	4	3.98
Phenylacetylene	—	5.71
I. Hydroxyl position[i]		
Methanol	(0)	0.01
Ethanol	1	1.00
Propanol	1	0.99
sec-Butanol	(2)	2.03
t-Butanol	3	3.00
Pentanol	1	1.00

Table 9.8   Continued

	Test	Predicted
3-Pentanol	2	2.03
Hexanol	1	0.98
2-Hexanol	(2)	2.08
2-Methyl-2-pentanol	3	2.99
3-Methyl-3-pentanol	3	2.98
Octanol	1	1.01
2-Propyn-1-ol	—	−2.06
J. Unity[i]		
Methanol	1	0.98
Ethanol	1	1.01
Propanol	1	0.99
sec-Butanol	1	1.00
t-Butanol	1	1.05
3-Pentanol	1	0.97
Hexanol	1	1.00
2-Methyl-2-pentanol	1	1.01
4-Heptanol	1	0.99
Octanol	1	1.03
Cyclohexanol	1	1.04
2-Propen-1-ol	1	0.99
2-Propyn-1-ol	1	1.04

[a] Reprinted with permission from D. G. Howery, P. H. Weiner, and J. S. Blinder, *J. Chromatogr. Sci.,* **12,** 366 (1974).

[b] Reprinted with permission from J. M. Soroka and D. G. Howery, unpublished work.

[c] Reprinted with permission from P. H. Weiner and J. F. Parcher, *Anal. Chem.,* **45,** 302 (1973).

[d] Reprinted with permission from P. H. Weiner and D. G. Howery, *Anal. Chem.,* **44,** 1189 (1972).

[e] Reprinted with permission from P. H. Weiner and D. G. Howery, *Can. J. Chem.,* **50,** 448 (1972).

[f] Reprinted with permission from R. B. Selzer and D. G. Howery, *J. Chromatogr.,* **115,** 139 (1975).

[g] Reprinted with permission from J. H. Kindsvater, P. H. Weiner, and T. J. Klingen, *Anal. Chem.,* **46,** 982, 1974.

[h] Reprinted with permission from D. G. Howery, *Anal. Chem.,* **46,** 829 (1974).

[i] Reprinted with permission from P. H. Weiner, C. Dack, and D. G. Howery, *J. Chromatogr.,* **69,** 249 (1972).

[j] Results are given for representative solutes. For complete results, consult the original articles. Test points in parentheses were known values free-floated on the test vectors; dashes indicate that points were free-floated.

solutes. Predicted values for the free-floated designees were used to devise an effective carbon-number scale for solutes having unsaturation, branching, and aromatic character. A scale for interrelating the effects of various kinds of carbon–carbon multiple bonds was developed by Howery[12] (see example H). On this scale saturated solutes were assigned a value of 0, solutes containing a double bond were assigned a value of 1, solutes containing a benzene ring were given a value of 3, and hydrocarbons containing a triple bond were free-floated. Styrene was assigned a test value of 4, since the contributions of aromaticity and double bonding were considered to be additive. The results of the test confirmed the general form of the vector and predicted an average value slightly greater than three for the triple bond on this scale. The additivity assumption was substantiated by free-floating phenylacetylene. That solute had a predicted value of 5.7, only slightly less than the value of six postulated for a molecule having a triple bond plus benzene ring.

9.8.3   Solvent Factors

Most GLC stationary-phase solvents are polymeric and often are poorly characterized. Physical-property data even for the common solvents are not available. Because it is extremely difficult to formulate structural vectors for complex polymers, nearly all of the insight into solvent cofactors has been obtained from studies involving relatively pure, well-defined monomeric solvents.

Four examples of solvent tests are given in Table 9.9. Examples A and B concern aqueous solvents. In example A, Kindsvater et al.[13] showed that the unity vector tested adequately for a group of mixed, liquid chromatographic solvents if only the first six, most polar solvents were included in the factor analysis (results given in the third column). Unity did not transform as a factor when

Table 9.9   **Examples of target tests of solvent vectors**

	Test	Predicted	
A. Unity[a]		6 Solvents	8 Solvents
$CH_3CN:H_2O$ (11:9)	1.0	0.94	1.12
$CH_3CN:H_2O$ (6:4)	1.0	0.93	0.81
$CH_3CN:H_2O$ (4:6)	1.0	1.04	1.39
$CH_3CN:H_2O$ (3:1)	1.0	1.19	0.99
$CH_3CN:H_2O$ (4:1)	1.0	0.98	0.90
$CH_3COCH_3:H_2O$ (1:1)	1.0	0.86	0.75
$CH_3OH:H_2O$ (12:1)	1.0	—	0.95
Dioxane:H_2O (3:1)	1.0	—	0.25

Table 9.9 Continued

	Test	Predicted	
B. Bulk volume per gram[b]		(1 factor)	(2 factors)
Packing 1[c]	0.29	0.13	0.30
Packing 2[c]	0.25	0.17	0.26
Packing 3[c]	0.20	0.18	0.19
Packing 4[c]	0.17	0.22	0.17
Packing 5[c]	0.14	0.25	0.15
C. Interaction parameter[d]			
Heptadecane	0.0	−0.1	
1-Hexadecene	1.6	1.8	
1-Hexadecyl chloride	5.8	6.0	
1-Hexadecyl bromide	6.5	6.5	
1-Hexadecyl iodide	7.1	7.1	
Dioctyl ether	4.3	4.1	
Dioctyl thioether	6.2	6.1	
D. McReynolds, Iodobutane constant[e]			
Diglycerol	245	248	
Diisodecyl phthalate	83	86	
Dioctyl sebacate	68	70	
Flexol 8N8	98	90	
Hallcomid M18	82	77	
Hyprose SP-80	310	287	
Isooctyldecyl adipate	(72)[f]	69	
Quadrol	208	233	
Sucroseoctaacetate	292	300	
TMP Tripelargonate	77	85	
Tricresol phosphate	169	159	
Zonyl E7	146	144	

[a] Reprinted with permission from J. H. Kindsvater, P. H. Weiner, and T. J. Klingen, *Anal. Chem.,* **46,** 982 (1974).

[b] Reprinted with permission from P. H. Weiner, H. L. Liao, and B. L. Karger, *Anal. Chem.,* **46,** 2182 (1974).

[c] Packings contain different loadings of aqueous tetraethylammonium bromide.

[d] Reprinted with permission from D. G. Howery and J. M. Soroka, unpublished work.

[e] Reprinted with permission from J. M. Soroka and D. G. Howery, unpublished work.

[f] Known value in parentheses was free-floated.

the last two mixed solvents, which have considerably lower dielectric constants, were incorporated in the data matrix (results in the fourth column). In example B, the volume of liquid per gram of packing was felt by Weiner et al.[11] to be a measure of the second most important solvent factor. In confirmation of theory, the vector tested well in a two-factor space (fourth column), whereas the transformation was unsuccessful for a one-factor model (third column).

Target testing has been used to corroborate empirical scales for solvents. For instance, Howery and Soroka,[19] in a study of well-characterized monomeric solvents, tested a scale designed to measure the van der Waals interaction strenghts of solvent functional groups. This parameter was calculated by Zielinski and Martire[27] from a least-squares analysis of retention indices. The results in example C confirmed that the interaction parameter is a basic solvent factor. In another problem involving monomeric solvents, the well-known McReynolds constants[24] were shown by TFA to be solvent factors. According to McReynolds's approach, a solvent is characterized by a set of 10 constants associated with 10 simple solute "probes." In example D, Howery and Soroka[20] verified that the McReynolds constant for iodobutane solute transformed as a solvent factor. The predicted constant for isooctyldecyl adipate, free-floated on the test vector, substantiated the test.

9.9 KEY SETS OF BASIC VECTORS

The best models resulting from combination-TFA of basic vectors are summarized in Table 9.10. Each vector that occurs in more than one solution is abbreviated with symbols defined in the footnote. Considering the complexity of chromatographic processes, several of the TFA models are strikingly good. Upon comparing the RMS errors resulting from the key combinations with the experimental errors (see Table 9.10), we find that especially satisfactory solutions have been formulated for solutes in two problems[13,15] and for solvents in two problems.[11,19] In the earlier combination studies, only a few selected sets of basic vectors were tested. By contrast, in several of the more recent studies, all possible combinations of 25 or more vectors involving as many as seven vectors per combination set have been examined.

Even in the earliest use of the combination procedure, Weiner et al.[6] found a set of solute vectors that accounted for the major factors. Kindsvater et al.[13] obtained a TFA model that reproduced liquid-chromatographic data within experimental error. As an example of thoroughness, Selzer and Howery[15] studied all combinations of 33 basic vectors in a six-factor space. Over 1 million combinations were required. They proposed the use of pattern tables (see Section 9.6) for identifying the more important cofactors and for pinpointing similar factors.

Table 9.10 Summary of key combinations of basic vectors

Reference	Error, TFA Solution[a]	Experimental Error[b]	Factors in Best Solution[c]
		Solutes	
6	8.4	5	BP, CN, MR, MW, UN
8	5	3	BP, either CN or MR, UN, van der Waals' *a* constant
12	8	2	Aromatic uniqueness, carbon-bond scale, CN, either enthalpy of vaporization or molecular volume, multiple-bond scale, triple-bond scale
13	0.04	0.03	CN, log GLC retention volume, UN
15	5.3	5	AN, CN, chain difference, chain ratio, square BP, either square CN or square AN or square (C + O) number
19	2.7	1	AN, CN, DM
20	9.8	5	CN, enthalpy of combustion, ketone uniqueness, MS, vapor pressure
21	11.0	5	Alcohol uniqueness, DM, MS, surface tension, total number of carbons, van der Waals' *b* constant
		Solvents	
11	0.5[d]	2[d]	Interfacial area/g, bulk volume/g
19	0.7	1	Log refractive index, retention index dispersion, UN
21	6.5	5	Correlation vector, DU, hydroxide uniqueness, McReynolds constant nitropropane, log sum McReynolds constants, Zonyl uniqueness
20	6.4	5	Correlation vector, DU, McReynolds constants for butanol and iodobutane, McReynolds *b* constant, UN

[a] Average or RMS error in r.i. units.
[b] Estimated upper limit of experimental error in r.i. units.
[c] Vectors appearing more than once in the table are abbreviated as follows: AN—atom number; BP—boiling point; CN—carbon number; DM—dipole moment; DU—diglycerol; MR—molar refraction; MS—magnetic susceptibility; MW—molecular weight; UN—unity.
[d] Percentage error.

In the key combinations for solutes in Table 9.10, carbon number appears seven times and unity appears three times. Structural vectors, including a number of clustering uniqueness tests, were incorporated in a majority of the key combinations. In most of the combination studies, several different models gave nearly equivalent data reproduction, implying the equivalence of some of the basic factors.

The most satisfying tie-ins with theory have been obtained for tests of solvent factors. Using TFA, Weiner et al.[11] verified, in detail, a theoretical model for retention mechanisms. For saturated alkane solutes, they substantiated a model involving only one factor, gas–liquid interfacial surface adsorption. When polar and unsaturated solutes were added to the data matrix, a second solvent factor, related to bulk gas–liquid partition (see Table 9.9, example B), was identified by target transformation, again consistent with the theoretical expectations. As a second example, Howery and Soroka[19] verified and extended a three-factor model for solute–solvent interactions proposed by Zielinski and Martire.[26] Several empirical parameters for both solutes and solvents were devised by Zielinski and Martire. Howery and Soroka corroborated these scales using target testing (see Table 9.9, example C). A thorough search for solvent factors led to a solution that reproduced the data within experimental error, as shown in Table 9.10. A moderately good solution was also formulated for the solute part of the same problem. The combination-TFA solution consisting of three solute factors and three solvent factors yielded a complete model which was theoretically reasonable. The three factors in the model consisted of a dispersion term, a dipole–induced dipole term, and a dipole–dipole term.

9.10 PREDICTIONS BASED ON KEY SETS OF BASIC VECTORS

For three GLC problems,[6,19,22] new rows for a data matrix have been predicted using sets of basic vectors (see Section 6.9). Since retentions for even simple alkane–alkane GLC systems have not been predicted theoretically, these results demonstrate the power of target factor analysis.

In the earliest application using basic vectors, Weiner et al.[6] predicted the retention indices of four new alcohols on 20 solvents. The five basic factors employed in the calculation are listed in Table 9.10. Data points were predicted with an average error of about 20 r.i. units, an encouraging result for an exploratory effort. In a second application, Selzer and Howery[22] predicted the retention indices of two new ethers on 20 solvents with an average error of only 3.5 r.i. units, well within the experimental error of 5 r.i. units.

A different approach was explored by Howery and Soroka.[19] Key sets of three basic factors were found for both the solutes and the solvents (see Section 9.9).

Accordingly, (9.1) was rewritten as

$$d_{ik} = k_1 u_{i1} v_{1k} + k_2 u_{i2} v_{2k} + k_3 u_{i3} v_{3k} \qquad (9.4)$$

Here k_m is the overall proportionality constant for the mth factor, and u_{im} and v_{mk} are the mth key basic factors for solute i and solvent k, respectively. Constant k_m is the product of the proportionality constants for the mth solute term and the mth solvent term. The basic factors used in the model are listed in Table 9.10. In view of (9.4), if values for three datum and for the corresponding three solute factors and three solvent factors are known, values for the three unknown k_m's could be calculated algebraically from three simultaneous equations. Using this approach, Howery and Soroka predicted r.i. data for new solute–solvent pairs for which values of the key factors were available from other sources. The errors in prediction for 11 new r.i. data averaged 10.3 r.i. units, thus confirming the validity of the overall TFA model.

9.11   SUMMARY OF RESULTS

Factor analytical studies of gas-chromatographic retention data have served as a major testing ground for applications of factor analysis in chemistry. Some preliminary conclusions from FA–GLC are listed below.

1. The number of factors predicted from factor analysis generally has been consistent with chemical insight.
2. Solvents have been classified using principal factors.
3. The uniqueness test has identified atypical molecules and has grouped molecules in clusters consistent with chemical insight.
4. Data matrices have been adequately reproduced using small sets of typical vectors.
5. New retention data have been predicted accurately using key sets of typical vectors.
6. A large number of physical and structural factors for solutes and solvents have been target tested successfully.
7. Molecular weight, heat of vaporization, and dipole moment appear to be solute factors in several problems.
8. Gas-phase imperfection, various structural vectors, and unity (for solutes having a common functional group) appear to be solute factors.
9. Complete TFA models involving basic vectors have been developed separately for solutes and for solvents.
10. An overall model involving pairs of solute and solvent factors has been formulated using combination TFA.
11. New retention data have been predicted adequately using key sets of basic vectors.

REFERENCES

1. B. L. Karger, L. R. Snyder, and C. Horvath, *An Introduction to Separation Science,* Wiley-Interscience, New York, 1973.
2. L. Rohrschneider, *J. Chromatogr.,* **22,** 6 (1966).
3. P. T. Funke, E. R. Malinowski, D. E. Martire, and L. Z. Pollara, *Sep. Sci.,* **1,** 661 (1966).
4. P. H. Weiner and D. G. Howery, *Can. J. Chem.,* **50,** 448 (1972).
5. P. H. Weiner and D. G. Howery, *Anal. Chem.,* **44,** 1189 (1972).
6. P. H. Weiner, C. Dack, and D. G. Howery, *J. Chromatogr.,* **69,** 249 (1972).
7. P. H. Weiner and J. F. Parcher, *J. Chromatogr. Sci.,* **10,** 612 (1972).
8. P. H. Weiner and J. F. Parcher, *Anal. Chem.,* **45,** 302 (1973).
9. S. Wold and K. Andersson, *J. Chromatogr.,* **80,** 43 (1973).
10. J. F. K. Huber, E. T. Alderlieste, H. Harren, and H. Poppe, *Anal. Chem.,* **45,** 1337 (1973).
11. P. H. Weiner, H. L. Liao, and B. L. Karger, *Anal. Chem.,* **46,** 2182 (1974).
12. D. G. Howery, Anal. Chem., **46,** 829 (1974).
13. J. H. Kindsvater, P. H. Weiner, and T. J. Klingen, *Anal. Chem.,* **46,** 982 (1974).
14. D. G. Howery, P. H. Weiner, and J. S. Blinder, *J. Chromatogr. Sci.,* **12,** 366 (1974).
15. R. B. Selzer and D. G. Howery, *J. Chromatogr.,* **115,** 139 (1975).
16. D. H. McCloskey and S. J. Hawkes, *J. Chromatogr. Sci.,* **13,** 1 (1975).
17. S. R. Lowry, G. L. Ritter, H. S. Woodruff, and T. L. Isenhour, *J. Chromatogr. Sci.,* **14,** 126 (1976).
18. M. Chastrette, *J. Chromatogr. Sci.,* **14,** 357 (1976).
19. D. G. Howery and J. M. Soroka, submitted for publication.
20. J. M. Soroka and D. G. Howery, submitted for publication.
21. D. G. Howery and J. M. Soroka, submitted for publication.
22. D. G. Howery, in B. R. Kowalski (Ed.), *Chemometrics: Theory and Applications,* ACS Symp. Ser. 52, American Chemical Society, Washington, D.C., 1977, p. 73.
23. W. O. McReynolds, *Gas Chromatographic Retention Data,* Preston Technical Abstract Co., Niles, Ill., 1966.
24. W. O. McReynolds, *J. Chromatogr. Sci.,* **8,** 685 (1970).
25. E. R. Malinowski, *Anal. Chem.,* **49,** 612 (1977).
26. R. W. Bach, E. Dotsch, H. A. Friedrichs, and L. Marx, *Chromatographia,* **4,** 459 (1971).
27. W. L. Zielinski and D. E. Martire, *Anal. Chem.,* **48,** 1111 (1976).

10 Additional Applications

The variety of applications discussed in this chapter further demonstrate the broad applicability of factor analysis. A summary of the kinds of problems covered in the chapter is tabulated in Table 10.1. In Sections 10.1 through 10.4 we focus attention on fundamental data which might be expected, theoretically or intuitively, to have factor analytical solutions. In Sections 10.5 through 10.7 we discuss some very complicated problems of a practical vein for which factor analysis also has furnished useful insights.

10.1 LINEAR FREE-ENERGY RELATIONSHIPS

Linear free-energy relationships (LFER) have received much attention during the past 30 years.[1] The LFER approach attempts to express reaction rates and equilibrium constants for a series of structurally related compounds in terms of properties of the substituent groups and the reaction media. The empirical Hammett equation,

$$\log K' = \log K^0 + \rho\sigma \tag{10.1}$$

Table 10.1   Summary of problem types discussed in this chapter

Field	Data Factor-Analyzed	Kinds of Designees
Linear free-energy relationships	Log acidity constant	Solutes, solvents
Mass spectrometry[a]	Spectral intensity	Molecules, m/e's
Instrument comparison	Absorbance	Solutions, wavelengths
Method comparison	Concentration	Methods, blood sera
Medium comparison	Absorbance	Compounds, wavenumbers
Solubility	Log solubility	Solutes, solvents
Solution properties	Thermodynamic parameter	Solutes, solvents
Polarography	Half-wave potential	Metal ions, solvents
Chelation	Log stability constant	Metal ions, ligands
Bond energy	Bond dissociation energy	Radicals, radicals
Biomedical chemistry	Biological parameters	Drugs, drug properties
Environmental chemistry	Concentration	Components, sites

[a] See also Chapter 7.

constitutes the origin of LFER. In (10.1), K^0 is the rate constant or equilibrium constant for the parent molecule, K' is the same quantity measured under identical experimental conditions for the parent molecule which has been modified by the presence of a substituent group, ρ is a reaction medium constant which is independent of the substituents, and σ is a constituent constant which is independent of the reaction medium.

Originally, the Hammett equation applied only to benzene compounds containing substituents in the meta and para positions, failing to account for ortho substituents. Higman[2] applied abstract FA to LFER in an attempt to account for the ortho substituents. Speculating that the Hammett equation might contain additional terms due to a variety of effects caused by substituent groups, he postulated that a more appropriate expression might be

$$\log K' = \log K^0 + \sum_i \rho_i \sigma_i \qquad (10.2)$$

where the sum is taken over all possible effects of the substituents, such as reaction field, polarization, inductive, and steric. Equation (10.2) is ideally suited for factor analysis. Unfortunately, Higman was unable to find a satisfactory solution due to the inadequacy of the graphical techniques he employed.

Malinowski[3] investigated the acidity constants of 13 ortho-, meta-, and para-substituted phenyl compounds in four reaction media. When the data were

Table 10.2 Target tests of solute test vectors for acidity data[a]

Solute	Unity		Hammett σ Constant		Covalent Radius	
	Test	Prediction	Test[b]	Prediction	Test	Prediction
$-$H	1	1.00	0.000	0.005	0.00	-0.18
o-F	1	1.04	—	0.690	0.72	0.87
m-F	1	0.93	0.337	0.339	0.00	0.23
p-F	1	1.00	0.062	0.059	0.00	0.07
o-Cl	1	1.07	—	0.823	0.99	0.97
m-Cl	1	1.03	0.373	0.373	0.00	-0.01
p-Cl	1	1.02	0.226	0.219	0.00	-0.03
o-Br	1	1.04	—	0.862	1.14	1.08
m-Br	1	1.00	0.391	0.374	0.00	0.02
p-Br	1	0.99	0.232	0.234	0.00	-0.05
o-I	1	0.90	—	0.920	1.33	1.23
m-I	1	0.98	0.352	0.371	0.00	0.04
p-I	1	0.98	0.276	0.276	0.00	-0.04

[a] Reprinted with permission from E. R. Malinowski, Ph.D. thesis, Stevens Institute of Technology, Hoboken, N.J., 1961.
[b] Dashes indicate free-floated test points.

subjected to factor analysis, three primary factors emerged. Two solute factors were target-tested: unity, to account for the unity coefficient multiplier of the log K^0 term, and Hammett's σ constant, obtained from previous studies involving only meta and para groups. The success of the two tests is seen in Table 10.2. Since σ constants for ortho groups did not exist, these points were free-floated, allowing the target test to yield predictions for these points. As seen in the table, σ values predicted for a substituent in the ortho position are much greater than those of the same substituent in the meta or para positions. At that time, these results were surprising, since the electronic effect of a substituent in the ortho position was generally assumed to differ little from that in the meta or para position. The third solute factor was suspected to be caused by steric hindrance of the substituent in the ortho position. To account for this factor, Malinowski developed a test vector related to the size of the substituent group, consisting of the covalent radii of the ortho substituents and of zeros for the meta and para substituents. This test also gave excellent correlation, as shown in Table 10.2.

The three solute factors, used in combination, reproduced the data within experimental error and therefore accounted for all the important factors of the solute space. Furthermore, the loading coefficients obtained from combination-TFA, which correspond to the solvent cofactors, ρ, compared favorably with the reaction medium constants calculated by Jaffe[4] from data involving meta- and para-substituted compounds only.

Weiner[5] factor-analyzed the acidities of 19 substituted benzoic acids in seven solvents. Four factors were required to reproduce the data within the reported experimental error. Uniqueness tests for the solutes showed that none of the solutes were atypical in behavior. However, when applied to the solvents, the uniqueness tests indicated that ethylene glycol possessed a unique characteristic. This characteristic could not be attributed to hydrogen bonding because the other alcohol solvents exhibited low predicted values on the test for ethylene glycol uniqueness. To account for the four factors, Weiner proposed the following model:

$$\log \frac{K(i, k)}{K^0} = \left(\log \frac{K(i, \text{gas})}{K^0} \right) \cdot 1 + U_E(i, j) \cdot V_E(j, k)$$

$$+ U_W(i, j) \cdot V_W(j, k) + U_G(i, j) \cdot V_G(j, \text{ethylene glycol}) \quad (10.3)$$

The first term on the right-hand side of this equation accounts for the acidity of the proton transfer reaction in the absence of solvent (i.e., in the gas phase) with the solvent cofactor for the gas-phase term being unity. The second term is the electrostatic contribution, the third term the van der Waal's (dispersion) effect, and the fourth term accounts for the unique effect produced by ethylene glycol solvent. The U's refer to solute cofactors and the V's refer to solvent cofactors.

Weiner used target testing to identify all four solute factors. Two of the solvent factors: unity (the solvent part of the gas-phase acidity term) and the uniqueness term for ethylene glycol, were easily target-tested because a theoretical model was not required. For the electrostatic contribution, Weiner utilized the theoretical model proposed by Kirkwood and Westheimer,[6] which expressed this term as a product function of solute and solvent terms, as required for factor analyzed. The solvent electrostatic factor was simply the reciprocal of the dielectric constant of the medium. For the van der Waals contribution, Weiner took advantage of the success of a similar term developed in NMR–FA studies (see Section 8.1.6), where the solvent part of the term was found to be a function of the polarizability and the electron-cloud distribution of the solvent molecule. Target transformation into each of these four solvent factors was successful, as shown in Table 10.3.

Simultaneous target transformation into the combination of the four solvent test factors yielded an RMS error about three times the experimental error in the acidity data. According to (10.3), the coefficients resulting from combination TFA for the solvent unity term should correspond to the solute gas-phase acidities. Thus gas-phase acidities were predicted solely from solution measurements. Unfortunately, no experimental data for such acidities existed at that time for comparison. Weiner checked the validity of these results by repeating

Table 10.3   Target tests of solvent vectors for acidity data[a]

Solvent	Unity		Reciprocal of Dielectric Constant		Van der Walls' Effect		Uniqueness for Ethylene Glycol	
	Test	Pre-diction	Test	Pre-diction	Test	Pre-diction	Test	Pre-diction
Methanol	1	1.02	3.17	2.89	7.44	7.41	0	0.00
Ethanol	1	1.00	4.14	4.98	6.93	6.78	0	−0.10
Ethylene glycol	1	0.96	2.66	2.95	7.59	7.59	1	0.93
Butanol	1	0.97	5.73	5.31	6.78	6.72	0	0.19
Propanol	1	1.03	4.98	4.66	6.85	7.08	0	0.00
Dioxane–water ($\epsilon = 55$)	1	1.00	1.82	1.29	7.84	7.84	0	0.00
Dioxane–water ($\epsilon = 40$)	1	1.00	2.50	3.36	7.72	7.68	0	0.00
Dioxane–water ($\epsilon = 15$)	1	1.00	6.68	6.23	7.46	7.49	0	0.00

[a] Reprinted with permission from P. H. Weiner, *J Am. Chem. Soc.,* **95,** 5845 (1973).

the factor analysis after deleting ethylene glycol from the data matrix. If the fourth factor were truely unique to ethylene glycol, the factor space would be reduced to three and the first three terms in (10.3) would suffice. When the unique solvent was removed, the same solute cofactors were obtained for the first three terms.

Wold and Sjostrom[7] used principle factor analysis (PFA) to determine which reaction series follow the simple Hammett relationship (10.1) and which reaction series require an expanded Hammett equation (10.2). As a secondary goal, they also used PFA to obtain the best set of σ values from all the series which were shown by PFA to obey the simple Hammett relationship.

In one calculation, the constant in the Hammett equation was set equal to the logarithm of the equilibrium constant of the unsubstituted parent molecule of the series, giving a "restricted" constant. In a second calculation, the constant was set equal to the mean value of the logarithm of the equilibrium constants for the series, giving an unrestricted constant. A complete data set and an incomplete data set involving only meta substituents with free electron pairs (donors) and p-NO_2 groups were factor-analyzed. The latter set was chosen to ensure the absence of resonance effects. Two conclusions were reached. First, the overall fit with the unrestricted constant was slightly better in all cases. Second, σ values generated by PFA for the unrestricted cases led to data reproduction that was far better than the obtained using previously published σ values.

10.2 MASS SPECTROMETRY

Mass spectra (MS) of large organic molecules exhibit hundreds of fragmentation lines containing detailed structural and mechanistic information. Because of the complexity of the ionization process and the myriad of uncontrolable variables, MS patterns are poorly understood.

Justice and Isenhour[8] believed that factor analysis could show how the functional groups of the parent molecule influence the mass spectral pattern. They carried out an ambitious study involving 630 low-resolution mass spectra of compounds having an elemental composition of the form $C_{2-10}H_{2-22}O_{0-4}N_{0-2}$. Their data matrix incorporated the intensities of 119 m/e positions ranging from 12 to 141 mass units. The experimental data were standardized by subtracting the mean intensity of the mass position and then dividing by the standard deviation of the mass position, a modified form of correlation about the mean. Retaining only those factors having eigenvalues greater than 1 (see Section 4.3.2), 42 factors, accounting for only 73% of the total variance, were deemed important. Attempts were made to relate the loading coefficients to the presence of functional groups such as carbon chain, phenyl, carbonyl, ether, hydroxyl, and amine. Ranking procedures were developed for this purpose and the factors were apportioned among the various functional groups.

A more selective approach was taken by Rozett and Petersen.[9-11] A series of detailed factor analytical studies of the mass spectral intensities of 22 benzenoid isomers having the formula $C_{10}H_{14}$ were carried out. They reasoned that since all these hydrocarbon isomers had the same molecular weight and contained a benzene ring, the factor space should be relatively simple, while the side-chain structures of the isomers should provide sufficient variety to study phenomena such as ring opening, ring expansion, and ion-neutral complexing, phenomena known to influence mass spectra.

Of the various factor analytical pretreatment methods studied, Rozett and Petersen[9] found that the best technique for MS utilized the absolute intensities in conjunction with covariance about the origin (see Section 3.2.3). With three principal factors, 99.1% of the variation in the data was accounted for and the data were reproduced within 1%, the average repeatability in measuring the height of a mass spectral line. For such a complicated problem, the factor size was surprisingly small. Combination TFA of various sets of isomers (typical rows) and fragments (typical columns) showed that there were a number of key sets of isomers and fragments which reproduced the data adequately.[10] Several key sets were found because some of the isomers and fragments were equally dependent upon the same real factor. Conversely, sets of isomers and sets of fragments that did not represent all three factors gave poor reproductions in combinations. Such studies helped classify the isomers and fragments into clusters.

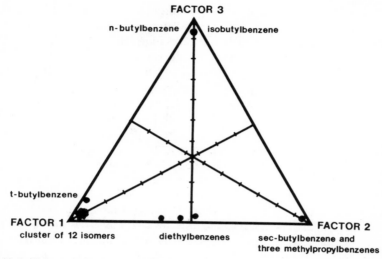

FACTOR 3

n-butylbenzene   isobutylbenzene

t-butylbenzene

FACTOR 1   FACTOR 2

cluster of 12 isomers   diethylbenzenes   sec-butylbenzene and
three methylpropylbenzenes

Fig. 10.1. Triangular plot of the square of the loadings obtained from varimax rotation of mass spectral data [Reprinted with permission from R. W. Rozett and E. M. Petersen, *Anal. Chem.,* **48,** 817 (1976)].

In order to examine the clusters that evolved from FA–MS, Rozett and Petersen[11] used triangular plots to represent the three-factor space. These plots were constructed so that the three triangular coordinates represented the three abstract factors. The sum of the squares of the three loadings for any specified isomer was standardized to equal 100, a mathematical requirement for any three-dimensional plot made on a two-dimensional plane. For the principal factor solution, the triangular plot exhibited three clusters of isomers. Sixteen, four, and two isomers clustered near the corners associated with the first, second, and third principal factors, respectively.

Because it is impossible to have negative intensities in mass spectra, the factor loadings obtained from PFA lacked theoretical interpretation. To give real significance to the loadings, Rozett and Petersen[10] used varimax rotation (see Section 3.4.2) to obtain a new set of abstract factors, called partial factors, all of which had positive loadings. Figure 10.1 is a triangular plot of the square of the loadings of the three factors obtained from the varimax rotation.[11] The clusters obtained after rotation were tighter and closer to the corners than the clusters in the principal factor plot. Furthermore, the three diethyl isomers formed a cluster midway along edge 1–2, and *t*-butylbenzene gave a unique point lying on edge 1–3. Clusters along the edges arose from linear combinations of the two connecting corner factors. Since no clusters appeared in the interior of the triangle, there were no isomers that depended upon all three rotated factors. In addition, such plots gave insights into the nature of the reaction mechanisms responsible for the mass spectral fragmentations. By studying the structures of the isomers in each cluster, Rozett and Petersen[11] concluded that corners 1, 2, and 3 represented fragmentation patterns indicative of a loss of one carbon atom, two carbon atoms, and a propyl group, respectively, from the parent ion. Edges 1–2 and 1–3 were attributed to a loss of both methyl and ethyl groups and to a loss of both methyl and propyl groups, respectively.

Encoded within the three abstract factors were three basic factors truly responsible for the fragmentation. In a search for the real factors, Rozett and Petersen[11] target-tested a variety of geometrical, physical, and thermodynamic properties. The geometric tests included structural features such as the Weiner path number (the sum of the number of bonds between all pairs of carbon atoms), the total number of side chains, the number of specific terminal groups (such as methyl, ethyl and propyl), and the number of ways of obtaining certain designated masses. The physical tests included melting point, boiling point, density, and refractive index. Thermodynamic tests involved properties such as the heat of formation, entropy of formation, heat capacity, and ionization potential. Although some of the individual tests were encouraging, no combination set of three basic vectors was found to be satisfactory. While all the tests involved parameters associated with the neutral parent molecule, it was deemed more likely that the true factors were associated with properties of the parent ion.

The success of their initial studies led Rozett and Petersen[12] to expand their data base to include 70 aromatic hydrocarbons, ranging from C_7H_8 to $C_{12}H_{18}$, and 151 mass peaks, ranging from 12 to 162. The intensities of each compound were normalized by dividing each intensity by the square root of the sum of the squares of the intensities of the compound. To conserve on computer time, they developed a direct factor analysis procedure[12] for extracting the principal eigenvectors up to a specified number, without requiring the complete solution. This method is valuable when dealing with large data matrices which might have hundreds of eigenvectors.

Five factors, accounting for 91% of the variation of the data, were deemed acceptable.[12] An abstract rotation called *direct quartimin* was used to group the molecules into clusters. Quartimin involves rotating the eigenvectors so as to minimize the sum of the fourth powers of the loading cofactors. By studying the dependence of each compound on the quartimin vectors, a physical interpretation of the underlying factors emerged. Factor 1 was associated with the presence of the $C_9H_{11}^+$ ion, factor 2 with $C_8H_9^+$, factor 3 with $C_7H_7^+$, factor 4 with $C_{10}H_{13}^+$, and factor 5 with $C_{11}H_{15}^+$.

Burgard and coworkers[13] factor-analyzed the mass spectra of oligodeoxyribonucleotides in order to determine the presence or absence of a given nucleoside, the relative number of different nucleosides, and the sequence of the nucleosides in the compound. The raw data matrix consisted of the intensities of 32 selected ions eminating from 32 nucleotides. The nucleotides were composed of various

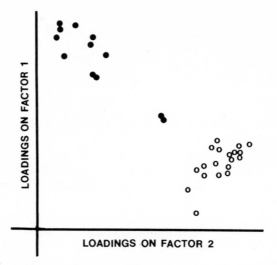

Fig. 10.2. Plot of the loadings on the first two principal axes resulting from analysis of the normalized-to-sum adenosine ions for compounds containing adenosine (filled circles) and compounds not containing adenosine (open circles) [Reprinted with permission from D. R. Burgard, S. P. Perone, and J. L. Wiebers, *Anal. Chem.*, **49**, 1444 (1977)].

amounts and various linkages of four nucleosides: adenosine, guanosine, thymidine, and cytidine.

Each data column, associated with a specific nucleotide, was normalized by dividing each ion intensity by the sum of the intensities of the 32 ions in the column. Principal factor analysis of the normalized data matrix, using covariance about the origin, yielded four eigenvectors which accounted for 99% of the variance. Although varimax rotation indicated that the varimax factors closely resembled the four nucleosides, the loadings could not be used to establish the presence or absence of a particular nucleoside in a compound.

To establish the existence of a particular nucleoside, the following procedure was adopted. The initial set of 32 ions was chosen on the basis of previous mass spectral studies so that a set of eight selected ions represented each nucleoside. To test for the presence of a specific nucleoside, each column of the data matrix was normalized by dividing each ion intensity by the sum of the intensities of the ions belonging to the subset associated with the nucleoside under consideration and principal factor analysis was carried out. The factor loadings on the first two principal axes obtained from testing for the presence of adenosine are plotted in Figure 10.2. The loadings of the 32 compounds were divided into two clusters separating the compounds which contained adenosine from those which did not. Similar results were achieved for the other three nucleosides.

To determine the relative number of different nucleosides in a compound, ratios of the intensities of selected ions for each pair of nucleosides were considered. The loadings from PFA gave a surprisingly good estimate of the relative amount of the two nucleosides present in a compound. Attempts to use factor analysis to obtain sequence information were unsuccessful.

10.3   COMPARISONS IN ANALYTICAL CHEMISTRY

Three types of factor analytical applications involving comparison methods in analytical chemistry are presented in this section. All these methods take advantage of the inherent statistics which is automatically built into the principal component feature of factor analysis.

Instrument Comparisons. Analytical chemists require methods to evaluate the performance of spectrophotometers. With this need in mind, Wernimont[14] examined the absorbance curves of a group of 16 spectrometers, all of the same model but spanning more than a 20-year manufacturing period. The absorbances of three solutions of potassium dichromate were measured at 20 wavelengths on two different days, yielding a 6 × 20 data matrix for each instrument. The rank of such a data matrix will be unity if the instrument is performing properly, whereas the rank will be two or more if the instrument is improperly adjusted, improperly calibrated or improperly read. The first factor arises from Beer's law for one component behaving ideally (see Chapter 7); additional factors arise from unique error factors such as wavelength shifts in the instrument or large reading errors by the chemist. The AFA methodology is simple to carry out on a routine basis. Regular application assures the spectroscopist that his instruments are performing properly and warns him when performance is unsatisfactory.

Using covariance about the origin, Wernimont calculated the residual standard deviations [see (4.44)] for each of the 16 data matrices, assuming rank one. Several instruments showed residuals that were much greater than 0.004 absorbance unit, the error estimated from individual measurements. In some cases close examination of the data revealed gross reading mistakes, which were easily corrected. More interestingly, six instruments still had unsatisfactory residual errors. After careful mechanical, electrical, and optical adjustments were made, all these erratic instruments produced acceptable spectra.

Method Comparison. New methods of analysis are a main pursuit of analytical chemists. Before replacing an old method with a new one, the chemist must evaluate the new technique for systematic and random errors, and for possible interfering substances which may be specific to the new method. Rec-

Table 10.4   Standard errors from PFA of six methods for measuring glucose in uremic sera [a,b]

Method	Model I		Model II		Model III	
	E_1	E_2	E_1	E_2	E_1	E_2
Ferricyanide	10.11	3.47	11.63	3.42	22.48	4.05
Neocuprine	7.06	5.37	6.96	5.41	11.53	5.35
o-Toluidine	4.21	4.16	4.93	4.24	4.67	4.57
Oxidase ABTS	6.76	2.64	99.50	2.69	22.56	3.06
Oxidase MBTH-DMA	3.11	2.15	4.26	2.14	4.14	2.92
Hexokinase	4.77	3.71	5.22	4.16	6.02	4.19

[a] E_1, RMS error assuming no interfering species; E_2, RMS error assuming one interfering species.
[b] Reprinted with permission from R. N. Carey, S. Wold, and J. O. Westgard, *Anal. Chem.*, **47**, 1824 (1975).

ognizing this problem, Carey et al.[15] applied principal factor analysis to method comparisons. PFA offers an advantage over older methods such as regression analysis by eliminating the need to choose a reference method which must be assumed to be free from errors and interferences. Instead, PFA allows the chemist to use a composite of different methods as a reference, thus involving multivariate statistics rather than univariate statistics.

Carey and coworkers investigated six methods for measuring glucose, listed in Table 10.4, involving 130 blood sera which were divided into uremic and nonuremic sera. Three PFA models were used to evaluate the data. Model I assumed that a systematic error C_i as well as a random error E_{ik} existed. The glucose value for the ith method applied to the kth sample, Y_{ik}, was expressed as

$$Y_{ik} = C_i + E_{ik} + \sum_{j=1}^{m} U_{ij}X_{jk} \qquad (10.4)$$

where U_{ij} was the response of the jth species and X_{jk} was the concentration of the jth species. The sum was taken over all m species which contribute to the measurement. If there were no interfering species, m would equal 1; if there were a single interfering specie, m would equal 2. Based on (10.4), PFAs were carried out and RMS errors in the reproduced glucose values were calculated for each method. These errors for uremic sera, labeled E_1 and E_2 for m equaling 1 and 2, respectively, are shown in Table 10.4. Model II assumed that the systematic error C_i was zero. Since the PFA results in Table 10.3 for uremic sera based on this model show that the E_2 values for models I and II do not differ significantly, the absence of systematic error was indicated. Model III assumed that the responses due to the first factor, U_{i1}, were equal to 1.00 for those methods cali-

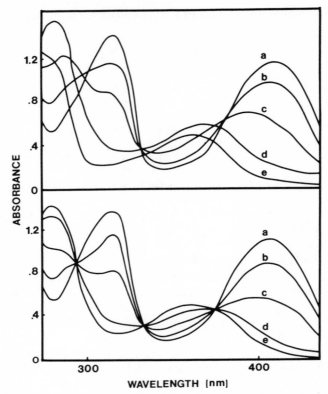

Fig. 10.3. Ultraviolet spectra of anthraquinone in aqueous sulfuric acid: (*a*) 99.0%, (*b*) 91.6%, (*c*) 86.3%, (*d*) 80.7%, and (*e*) 73.0%. Raw data curves are shown in the upper block. Reconstituted curves, obtained from the mean curve and first principal factor, are shown in the lower block [Reprinted with permission from J. T. Edward and S. C. Wong, *J. Am. Chem. Soc.,* **99,** 4229 (1977)].

brated by aqueous standards (ferricyanide, *o*-toluidine, and hexokinase), thus allowing for calibration constants which depended on the method employed.

By comparing E_1 to E_2 for a given method, we can conclude quickly which methods are sensitive to interfering substances. If, for a given method, E_2 approximately equals E_1, no interfering species are present. If E_2 is significantly smaller than E_1, an interfering specie is present. From all three models the same conclusions were reached. For both uremic sera and nonuremic sera, the ferricyanide and oxidase ABTS methods were found to be extremely sensitive to interfering substances; neocuprine and oxidase MBTH–DMA methods showed a moderate dependence upon foreign species; and the hexokinase and *o*-toluidine methods were relatively free from interferences Thus factor analysis can reveal which glucose measurements are subject to interference from foreign substances or to imprecision. Such information is a valuable aid in selecting the most valid analytical procedure to be used in routine analysis.

Medium Comparisons. Correcting analytical measurements for medium effects has been a vexing problem for the chemist. In many situations factor analysis may afford the most powerful method for identifying and compensating for medium effects efficiently and inexpensively.

An example of the use of factor analysis for this purpose is found in the work of Edward and Wong[16] concerning measurement of the ionization of carbonyl compounds in sulfuric acid. The ultraviolet spectra of 16 carbonyl compounds (involving aldehydes, ketones, and amides) in various concentrations of sulfuric acid did not exhibit isosbestic points characteristic of protonation equilibria. Figure 10.3, concerning anthraquinone, shows a typical result. The spectra (top set of curves) were digitized and the absorbance matrix was subjected to PFA. The first principal eigenvector, accounting for 96% of the variance, was associated with the effect of protonation. The second eigenvector, accounting for 3% of the variance, was associated with the medium effect. The remaining 1% variability was due to experimental error. Using only the first principal eigenvector, the PFA reconstituted curves exhibit clear isosbestic points as shown in the lower set of curves in Figure 10.3. The ionization ratios and equilibrium constants for the 16 carbonyl compounds were then obtained from the reconstructed spectra.

10.4 ADDITIONAL FUNDAMENTAL APPLICATIONS

Factor analyses of five kinds of physicochemical data are summarized in this section.

Solubility. Understanding the factors that influence the solubility of gaseous solutes in liquid solvents is one of the most fundamental problems in physical chemistry. Furthermore, the prediction of gas–liquid solubilities is of practical concern in chemical engineering. Factor analysis of solubility data has been employed for both of these purposes.

Howery and Chan[17] target-factor-analyzed a data matrix involving eight, nonpolar solutes (ranging from helium through oxygen to ethane) and 11 polar and nonpolar solvents (eight alkyl and aromatic hydrocarbons, chlorobenzene, ethanol, and dimethylsulfoxide). They analyzed the logarithm of the mole-fraction solubility because, from thermodynamics, the free energy of transfer of a gaseous solute into solution was known to be proportional to the logarithm of solubility, rather than to the solubility itself. Theoretically, the free energy was suspected to be a linear sum of terms related to the solute–solvent interactions responsible for the data. Howery and Chan were interested in identifying those solute and solvent basic factors that best model the solute–solvent interactions.

Four factors reproduced the solubility matrix within 5%, the upper limit of the experimental error. The factor indicator function (see Section 4.3.2) reached a minimum at four factors, further confirming the factor size. Uniqueness tests showed that ethane (uniqueness value = 0.97) and methane were the most unique solutes and that dimethylsulfoxide (uniqueness value = 0.86) and hexane were the most unique solvents. Combination TFA were used to determine which sets of typical factors (data vectors from the data matrix) best represented the data. For both key combination sets, the RMS error for the reproduction was within experimental error. The key set of typical solute vectors was associated with neon, argon, methane, and ethane, while the key set of typical solvent vectors involved hexane, heptane, benzene, and dimethylsulfoxide. The most unique designees (see above) were included in the key sets, a result usually noted in combination FA.

To identify parameters associated with the solute–solvent interactions, Howery and Chan[17] target-tested a large number of solute and solvent parameters, several of which had been suggested from previous theoretical and empirical studies of solubility. In the past, plots of solubility versus some property of the substances have been used to identify factors. Since such plots are expected to be linear only in one-factor problems, target testing affords a safer method for finding factors in multidimensional solubility problems. Solute factors that tested adequately included molecular mass, entropy of vaporization, entropy of solution, Henry's law constant, polarizability, hard-sphere diameter (results shown in Table 10.5), and Lennard-Jones force constant. Among the solvent parameters identified as basic factors were surface tension, log vapor pressure, rigid-sphere diameter, and a nonpolar interaction term. The well-known Hildebrand solubility parameter did not test very well. Including the functional forms of the test vectors, over 40 solute factors and nearly 35 solvent vectors tested moderately well or better.

In order to find which sets of parameters represented the best complete model for the solute and solvent parts of the interaction space, combination TFA of the basic factors was carried out. For both the solutes and the solvents, the key combination sets gave RMS errors less than twice the experimental error. Henry's law constant and polarizability were represented in nearly all of the best combinations of basic solute vectors. Surface tension seemed to be an especially important solvent factor.

A simple target factor analytical procedure for predicting new solubility data was demonstrated by Howery and Chan.[17] Their method involved target testing incomplete vectors of solubility data, thereby predicting solubilities for the

Table 10.5   **Target tests of solute test vectors for solubility data (Howery and Chan[17])**

Solute	Hard-Sphere Diameter		Log Solubility in Cyclohexanol	
	Test	Predicted	Test[a]	Predicted
He	2.63	2.62	−0.325	−0.327
Ne	2.78	2.79	—	−0.168
Ar	3.40	3.73	−0.724	−0.717
H_2	2.87	2.73	0.228	0.238
N_2	3.70	3.85	0.424	0.424
O_2	3.46	3.36	—	0.712
CH_4	3.70	3.65	1.099	1.100
C_2H_6	4.38	4.35	1.914	1.914

[a] Dashes indicate free-floated test points.

free-floated molecules. A typical result is shown in Table 10.5. The accuracy of such predictions depended upon the extent to which the test points spanned the factor space. Predictions were more reliable if test points for unique molecules and for molecules represented in the key sets of typical vectors were incorporated in the test vector.

In another effort to develop a practical method for predicting solubility data, deLigny and coworkers[18] carried out a combined abstract factor analysis/ multiple regression analysis on the free energy of solution and the entropy of solution for 20 gases in 39 solvents. The original paper should be consulted for details concerning the data sets. An iterative regression procedure was developed to predict data missing from the data matrices. Abstract factor analysis was then applied to the completed data matrices. Two factors accounted for the main features of both the free-energy and the entropy data. The method was tested by seeing how well measured values were predicted from the abstract cofactors. The standard deviations for log solubility and entropy of solution were 0.13 and 0.84 cal mol^{-1} deg^{-1}, respectively. For missing data, the accuracy of predicted values for log solubility was estimated to be 0.20. The approach predicted adequately several data points that were published in the literature after the predictive equations were developed.

Solution Properties. In order to better understand physicochemical properties of solutions, Fawcett and Krygowski[19] applied factor analysis to several sets of thermodynamic data. Properties such as enthalpies of solution, free energies of solution, enthalpies of transfer, and free energies of transfer were factor-analyzed. Included in the solute–solvent data matrices were both polar organic molecules and inorganic electrolytes. For each property, two principal eigen-

Table 10.6   Target test of ionic charge for polarographic half-wave potentials[a]

Ion	Test[b]	Predicted
Li^+	—	2.03
Na^+	1.0	1.16
K^+	1.0	1.15
Rb^+	1.0	0.92
Cs^+	1.0	0.87
Mg^{2+}	2.0	2.17
Ca^{2+}	2.0	2.06
Sr^{2+}	2.0	1.76
Ba^{2+}	2.0	1.78

[a] Reprinted with permission from D. G. Howery, *Bull. Chem. Soc. Jap.*, **45**, 2643 (1972).
[b] Test value for Li^+ was free-floated.

vectors, resulting from covariance about the mean, explained more than 90% of the variance in the property. In each problem, the two principal factors correlated linearly with empirical estimates of solvent acidity and solvent basicity.

Polarography. A study by Howery[20] of the polarographic half-wave potentials of five alkaline and four alkaline-earth ions in five polar solvents illustrates the application of target factor analysis in an area lacking a theoretical foundation. The objective was to build an empirical model for ion–solvent interactions. With three abstract factors, over 80% of the data·was predicted within 30 mV. The factor space seemed reasonably spanned since experimental errors exceeding 30 mV were suspected for only a few of the half-wave potentials.

Howery tested several solvent and ionic parameters by target transformation. Molar enthalpy of vaporization, donor number, and a radial correction term· tested moderately well as solvent vectors. A model based on these three solvent factors gave in combination TFA an average error of 28 mV, a quite good solution to the solvent part of the problem. Ionic charge, reciprocal of the crystalline ionic radius, and the ratio of ionic charge to crystalline ionic radius transformed well as ionic vectors. Results from target testing the ionic charge are shown in Table 10.6. The value for lithium ion in the test was deliberately free-floated, since lithium ion was known to produce high overvoltages in polar solvents. The predicted value of 2.03 implied that the lithium ion behaved like a divalent ion in this polarographic investigation.

Chelate Stability. Realizing that a detailed understanding of the factors affecting chelate stability might lead to guidelines for improving the selectivity of chelate reactions, Duewar and Freiser[21] conducted an abstract factor analysis on the logarithm of the formation constants for 14 diaminetetraacetic-acid ligands and 24 metal ions. Factor analysis is well suited to the study of complex ions, furnishing solutions involving factors for the ligands and for the central metallic ion. The logarithm pretreatment was dictated by the well-known proportionality between standard free-energy change and the logarithm of the equilibrium constant.

Four factors accounted for the data within experimental error. To facilitate physical interpretation of the unrotated abstract factors, a simplified three-factor model consisting of an average stability constant, a metal ion factor, and a ligand factor was invoked. The average stability constant corresponded to the first principal eigenvector, the ion factor correlated with the ratio of charge to ionic radius, but the ligand factor could not be identified. Contrary to expectations, the logarithm of the proton association constant for the ligands did not appear to be equivalent to any of the principal factors.

Bond Energy. Bond dissociation energies are among the most fundamental data in chemistry. A bond dissociation energy matrix can be formed by using radicals as both row and column designees, the radicals constituting the chemical groups being joined together by the bond in question. In an effort to study the factors influencing bond formation, Howery and Rubinstein[22] target-factor-analyzed a symmetrical 7 × 7 matrix of bond dissociation energies of some simple molecules.

Three factors reproduced 96% of the data within the estimated experimental error of 2 kcal. According to the results of uniqueness tests, hydrogen and phenyl were the most atypical radicals. Combinations of typical vectors indicated that hydrogen, isopropyl, and benzyl were the three key radicals, giving an RMS error in combination-TFA of 0.78 kcal.

Over 50 basic vectors and three functional forms of each basic vector were target-tested. The results indicated that more than 25 test factors were acceptable representations of basic factors. This large number of acceptable factors indicated that many of the factors were interrelated (i.e., were linear functions of each other). A combination-TFA model consisting of the radical-methyl stretching frequency, the radical-hydrogen bond length, and the diagonal of the data matrix produced an RMS error of 0.86 kcal, well within experimental error.

10.5 BIOMEDICAL CHEMISTRY

A variety of applications of factor analysis in biomedical chemistry have been published. Because of the inherent complexity of biomedical data and difficulty in obtaining reproducible, quantitative measurements, these problems are harder to deal with than those discussed in the previous sections of this chapter.

In an early study, Woodbury et al.[23] demonstrated that AFA could be used to predict biochemical data. Several points in the data matrix were deliberately omitted and then were predicted using the following iterative procedure. Arbitrary values were assigned to the missing points. New estimates for the missing points were predicted from the reproduction step of AFA. The new estimates were substituted in the data matrix, reproduction was carried out again, and a second set of estimated values was obtained. This process was repeated until the deleted data were predicted adequately. For a sulfa drug–tissue localization matrix, the predictions were moderately good. For a bacteria–antibody matrix, the predictions were rather poor. A drawback of the approach is the difficulty in determining the proper number of factors to use in the reproduction. In fact, Swain and coworkers[24] have shown that iterative factor analysis can lead to absurd results when points are missing in the data matrix.

Principal factor analysis was used to Sneath[25] to study the relationship between the chemical structure and the biological activity of 20 amino acids. The data matrix was a 20 × 20 "resemblance" table constructed statistically to take into account 134 attributes of the amino acids. Typical attributes incorporated in the resemblance table included solubility, optical rotation, chromatographic retention, presence or absence of various functional groups, moments of inertia, and the number of lone pairs of electrons. The same amino acids represented both the row designees and the column designees of the data matrix Each entry in the data matrix was generated by means of a correlation coefficient designed to measure the overall resemblance between each pair of amino acids. Since each amino acid exactly resembles itself, the diagonal elements of the matrix were unities. A resemblance value of zero indicated that the two amino acids had no similarity whatsoever; negative resemblance values indicated opposite characteristics.

Principal factor analysis of the resemblance matrix yielded four eigenvectors with eigenvalues greater than 1, together accounting for 69% of the total variation. By studying the unrotated factor loadings, Sneath related the four factors to the aliphatic character, degree of hydrogenation, aromatic character, and degree of hydrothiolation (involving the abilities of hydroxyl and sulfhydryl groups to form hydrogen bonds), respectively. These four factors were correlated with biological activity and then used to predict the activities of new peptides. The predictions, although not highly accurate, were better than those obtained by chance guessing.

Factor analysis has been successfully used as a preprocessing method for pattern recognition studies involving biological compounds. Chemical pattern recognition techniques, such as the "linear learning machines," attempt to classify a set of compounds into one of two specified categories. For example, an investigator may wish to predict whether or not a compound has a certain biological activity. In order to achieve such a binary classification, a set of linear descriptors representing chemical or physical properties must be found. De-

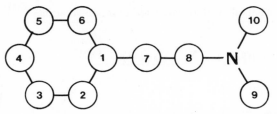

Fig. 10.4. Superstructure for 13 amino compounds [Reprinted with permission from A. Cammarata and G. K. Menon, *J. Med. Chem.,* **19,** 739 (1976)].

scriptors are continually added to the pattern recognition scheme until the classification reaches an acceptable degree of correctness. Ritter and Woodruff[26] pointed out that the number of descriptors employed does not necessarily equal the dimensionality of the factor space and that, in practice, often too many descriptors are used. By factor-analyzing a data matrix composed of the descriptors, the true size of the factor space and hence the minimum number of descriptors required can be determined. Thus factor analysis can play a valuable role in pattern recognition studies.

Pattern recognition factor analysis has been used to separate chemical compounds into therapeutic classes on the basis of their molecular structures. The method requires a general molecular descriptor coding which is applicable to the entire set of compounds. The coding must be designed to discriminate between atoms and groups. From the general descriptor a structural feature matrix is generated. Factor analysis of this structural feature matrix yields a minimum set of eigenvectors which can be used for pattern recognition classification, thus reducing the number of required axes to a minimum, simplifying both the classification and its interpretation. A reduction in axes is expected if the structural features are interrelated.

As an example of this technique, Cammarata and Menon[27] coded 13 amino compounds by means of a structural diagram, called a "superstructure," shown in Figure 10.4. Each arbitrary number in this diagram was used to designate a column in the molecule–feature data matrix. The descriptors shown in the footnote of Table 10.7 were proposed to distinguish between the different features of the molecules. The use of 1 or 0 served to indicate the presence or absence of a particular functional group. The use of 2 served to distinguish an aromatic carbon from an aliphatic carbon. Applying the feature descriptors to each of the 13 molecules, the molecule–feature matrix (Table 10.7) was generated. One can easily verify that the thirteenth row of Table 10.7 is a descriptor for benzylamine, $C_6H_5CH_2NH_2$.

Factor analysis of the resulting 10×10 correlation matrix yielded two eigenvalues greater than 1, accounting for 79% of the variance. A plot of the row-designee cofactors in the two-factor space is shown in Figure 10.5 (which

Table 10.7 Molecule-feature data matrix for pressor agents[a]

Compound	Feature[b]									
	X_1	X_2	X_3	X_4	X_5	X_6	X_7	X_8	X_9	X_{10}
1	2	2	2	2	2	2	1	1	0	1
2	1	1	1	1	1	1	1	1	0	1
3	1	1	1	0	1	1	1	1	0	1
4	2	2	2	2	2	2	1	1	1	1
5	1	1	1	0	1	1	0	1	0	0
6	1	1	1	0	0	0	0	1	0	0
7	2	2	2	2	2	2	1	1	0	0
8	1	1	1	0	0	1	0	1	0	0
9	1	1	1	0	0	0	1	1	0	0
10	2	2	2	2	2	2	1	0	0	0
11	2	2	2	2	2	2	0	1	0	1
12	2	2	2	2	2	2	0	0	0	1
13	2	2	2	2	2	2	0	0	0	0

[a] Reprinted with permission from A. Cammarata and G. K. Menon, *J. Med. Chem.*, **19**, 739 (1976).

[b] Based on the following descriptor set:

Feature	Nature	Descriptor
1–6	Aromatic atom	2
	Aliphatic atom	1
	No atom	0
7	CH_2 present	1
	CH_2 absent	0
8	$CHCH_3$	1
	CH_2	0
9, 10	CH_3 present	1
	CH_3 absent	0

is analogous to Figure 5.1). Each point in the plot is associated with a particular amino compound. The 13 compounds consisted of pressor agents (i.e., agents that cause an increase in blood pressure). Molecules known to be "strong" and "weak" pressor agents are labeled with solid circles and open circles, respectively; the classifications of points labeled with an X were not known previously. On the diagram the molecules fall into two distinct classes. When matched against known biological responses, the classifications based on factor analysis appeared to be correct.

In another classification study involving 43 compounds, Menon and Cammarata[28] employed coding values based upon atom and bond molar refractivities.

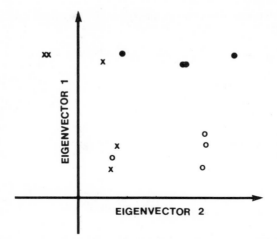

Fig. 10.5. Two-dimensional factor space showing weak (open circles), strong (closed circles), and undesignated (crosses) pressor agents [Reprinted with permission from A. Cammarata and G. K. Menon, *J. Med. Chem.,* **19,** 739 (1976)].

This coding was designed to take into account differences in "bioisosterism" among the functional groups, rather than simply account for the presence or absence of a functional group as in the previous example. Bioisosterism refers to those atoms, ions, or groups which have identical peripheral layers of electrons. The compounds included antihistamines, antidepressants, antipsychotics, anticholinergics, analgesics, and anti-Parkinsonian agents. The 43 × 8 pharmacophore–feature data matrix yielded four eigenvalues greater than 1, which accounted for 79% of the variance. Although the four-dimensional factor space could not be plotted, projections onto three-dimensional subspaces showed clusterings of points which were associated with various kinds of therapeutic activity.

To further simplify pattern recognition studies, Menon and Cammarata[28] used factor analysis in a two-step procedure. In the first step, factor analysis was used to separate an entire set of 39 drugs into broad clusters. Submatrices based on molecules in the smaller clusters were then factor-analyzed separately. Because the small clusters required fewer discriminators, the results were easier to interpret.

Weiner and Weiner[29] used target factor analysis to examine the structure–activity relationships of 16 diphenylaminophenol drugs characterized by 11 biological tests on mice. Eight factors were deemed necessary to reproduce the data satisfactorily. The factor-size decision was somewhat arbitrary, since several of the biological tests were qualitative in nature. Uniqueness values and clusterings based on the results of uniqueness tests were obtained for the drugs and,

Table 10.8   Uniqueness test for ring linkage to nitrogen atom in drugs[a]

Drug Substructure[b]	Test[c]	Predicted
$CH_2CH(CH_3)NCH_3$	0.0	0.01
$CH(CH_3)CH_2N(CH(CH_3)_2)_2$	0.0	0.00
$CH_2CH(CH_3)N(CH_3)CH_2C_6H_5$	(0.0)	0.09
$CH(CH_2CH_3)CH_2N(CH_2)_3CH_2$	1.0	0.91
$CH_2CH(CH_3)CH_2N(CH_2)_3CH_2$	(1.0)	1.00
$CH(C_6H_5)CH_2N(CH_2)_3CH_2$	1.0	1.05
$CH_2(CH_2)_3N(CH_2)_3CH_2$	(1.0)	1.0
$CHCH_2NCH_2CH_2CHCH_2CH_2$	—	1.16

[a] Reprinted with permission from M. L. Weiner and P. H. Weiner, *J. Med. Chem.*, **16**, 655 (1973).
[b] Structural component $ANRR'$ of the complete molecule $(C_6H_5)_2C(OH)ANRR'$ · HX.
[c] Known points in parentheses; dashed test value was free-floated.

what was even more intriguing, for the biological tests. Interrelating one biological test with key sets of others was also achieved.

Several structural vectors and a few vectors based on the results of the uniqueness tests were target-tested. For example, the presence of a ring linked to the nitrogen atom in the drug structure was shown to be a factor. Selected results for this test are shown in Table 10.8. A binary classification, separating the drugs into those having the ring linkage and those not having the linkage, was indicated. The large value predicted for the last drug appears to be related to the double ring in that compound. An attempt to predict the activity of new drugs using combination TFA based on a key set of eight typical biological-test vectors was not very successful. Little loss in predictive ability of the known drugs was observed when the ring-linkage factor shown in Table 10.8 was substituted for one of the key biological-test vectors.

In addition to the problems discussed above, applications of AFA to medicinal chemistry include studies of survival times for patients having cirrhosis of the liver using clinical and biological tests,[30] electroencephalograms of patients given or not given a drug,[31] effects of drugs on the membrane potential of brain cells,[32] and formulations of drug tablets.[33] Data matrices involving patients versus diagnostic tests and drugs versus evaluative tests are often obtained in medical and pharmaceutical research. Factor analysis should furnish useful classifications in such studies.

Prospective users of factor analysis in biomedical research should apply the technique with caution. Since many kinds of medical data may not obey the factor analytical model, the criteria for factor analyzability described in Chapter 4 should always be met. Malinowski,[34] using the theory of errors described in Section 4.3.2, found that the matrix employed by Weiner and Weiner[29] did not appear to have a good factor analytic solution. Howery and Gelbfish[35] investigated a data matrix concerning the concentrations of several components in the blood sera of patients having heart problems. Since no factor compression was observed, factor analysis was deemed inappropriate in that problem.

10.6 ENVIRONMENTAL CHEMISTRY

The ability of factor analysis to correlate environmental data has been established. Data matrices typically involve the analytical concentrations of several chemical components at either different sampling stations or sampling times. The objectives are to identify the sources of the components and ultimately to develop a detailed model for the processes responsible for generation and removal of components. Several representative applications are discussed in this section.

Problems involving elemental analysis of environmental samples are particularly amenable to factor analysis. In such problems the total amount of the ith element in the kth sample, x_{ik}, can be expressed as

$$x_{ik} = \sum_{j=1}^{n} c_{ij}q_{jk} \qquad (10.5)$$

Here c_{ij} is the relative concentration of element i in source j, q_{jk} is the amount of source j in the kth sample, and the sum is taken over the n significant sources of the element. Each source acts as a factor; summing over all sources gives the total amount of the element in the sample. The form of (10.5), involving a linear sum of products, indicates that this kind of problem should have a factor analytical solution.

Alpert and Hopke[36] tested (10.5) using artificial data sets. Twenty samples were prepared from known masses of five standard materials having known elemental compositions (the sources). Analyses of 37 elements were obtained for each sample using neutron activation analysis. Four criteria were employed to estimate the number of principal factors. The relative magnitude of eigenvalues, the Exner function (see Section 4.3.2), and the factor indicator function each indicated five factors, consistent with the number of sources used to prepare the samples, but the chi-squared test (see Section 4.3.1) gave no indication of the number of factors.

Alpert and Hopke[36] felt that target testing should identify the sources, since in the artificial data set the known elemental compositions of each source serve as test vectors. The results of target tests showed that the agreement between test and predicted values was good for elements present in high concentrations, but poor for trace components, which often tend to distinguish sources.

Factor analysis has been used in conjunction with cluster analysis in several studies of environmental problems. In "hierarchical" cluster analysis, each column designee is represented as a point in a c-dimensional space, where c is the number of columns in the data matrix. A variety of measures of similarity or dissimilarity are then used to classify the designees. The data points are assumed initially to form c separate, single-member clusters. Criteria such as nearest neighbors, farthest neighbors, and average-between-cluster distances are used to form the clusters in a stepwise procedure. The two points calculated to be most similar are combined into the first two-member cluster, which is then treated as a single point in the second stage of comparison with the $c - 1$ remaining points. The two most similar points in the new model are then merged to form a second cluster and the comparison calculation is repeated. At each stage of clustering, the two most similar clusters are combined, irrespective of the number of members in the merged clusters. Ultimately, a single cluster containing all c data points is formed. The connection diagram identifying the data points in each cluster at each stage of clustering is called a *dendogram*. The various clusters calculated from cluster analysis can be compared to those identified via principal or rotated factors.

A study of lake sediments by Hopke[37] illustrates the dual factor analysis/ cluster analysis approach. A data matrix representing 32 measurements of 79 sediment samples taken from sites throughout Chautauqua Lake, New York, was analyzed. The measurements included 15 elemental analyses, 12 properties related to grain size, water depth above the sample, and the percentages of sand, silt, clay, and organic matter. Both factor analysis and cluster analysis furnished insights into the mechanisms responsible for the distribution of properties. Five principal factors best described the data. Analysis of the loadings in the principal factors and rotated factors (based on a maximum likelihood method) led to assignments for the processes acting on the sediments. The five factors were related to the coarse-grain source material, the available surface area of the sediments, the glacial till source material, the active sediments due to stream deltas, and wave and current action, respectively. Hierarchical cluster analysis was used to cluster the sample sites. The two largest clusters of sites were shown to represent sites in the center of the lake and sites near shore. The near-shore sites were broken into three subclusters, the first being very sandy and relatively inactive, the second having active sedimentation and high wave action, and the third having intermediate activity.

Hopke et al.[38] were concerned with the concentrations of 18 chemical elements in air particulates collected at eight sampling stations in the Boston area. Six principal factors accounted for nearly 99% of the variation in the data. Clusterings of elements were identified following varimax rotation. From the clusters, the factors could be associated with various sources of particulate material. The first four factors were identified with crustal weathering dust, sea-salt aerosol, combustion of residual fuel, and automotive exhaust, respectively. Hierarchical cluster analysis was employed to further clarify the nature of the factors. Similarities among sites were detected by calculating factor scores for the larger clusters via a special procedure. Several clusters were shown to be strongly associated with particular factors. For example, a cluster that consisted almost entirely of downtown Boston sites was found to have a large score on the factor related by factor analysis to automotive exhaust.

The concentrations of 24 chemical species present in particulate matter collected at 11 locations in the Tucson, Arizona, area at various sampling times over a 1-year period were factor-analyzed by Gaarenstroom and coworkers.[39] Species-time data matrices for each location were factor-analyzed separately. Between five and eight factors were considered adequate to reproduce the data for the various sampling locations. By studying the loadings in the varimax-rotated solution, Gaarenstroom et al. attributed the cause of the pollutants to three main sources: soil, nonlocal aerosol, and automotive exhaust.

An abstract factor analysis of Puget Sound, Washington, rainwater, involving concentrations of 16 ions at 22 sampling stations, was conducted by Knudson et al.[40] Three factors accounted for about 70% of the variance in the data. A sea-salt background, a generalized urban source, and an industrial source were interpreted, by use of varimax rotation, to be the three main sources of the ions.

Several other factor analyses of environmental data have been published. Factor analytical investigations of large-scale pollution by Blifford and Meaker,[41] of the effects of meteorological parameters on air pollution by Peterson,[42] of the effects of air pollution on mortality by Lave and Seskin,[43] and of the sources of urban dust by Linton et al.[44] illustrate the scope of factor analysis environmental problems. Considering the growing need to understand the mechanisms that determine the distributions of pollutants, factor analysis of environmental data should be of even greater value in the future.

10.7 ADDITIONAL PRACTICAL APPLICATIONS

A hodgepodge of additional practical applications of factor analysis are referenced in this section.

In chemical industry, abstract factor analysis has been used to analyze a variety of data.[45] For example, in the food and beverage industries, correlations between the properties of rice and the flavor of Japanese sake,[46] between the quality of malt and of beer,[47] between the amino acid composition and varieties of potatoes,[48] and between the amino acid composition and varieties of barley[49] have been determined with factor analysis. Studies of geological data,[50] of the composition of moon rocks,[51] of interacting species in biological systems,[52] of taxonomic classifications of plants,[53] and of anaerobic digestion in ecological problems[54] give but a hint of the potential of factor analysis in sciences closely related to chemistry.

The sheer magnitude of the number of monographs on abstract factor analysis listed in the Bibliography testifies to the leading role of factor analysis in the behavioral sciences. Such is the importance of factor analysis in the social sciences that the method is referred to as the calculus of the social sciences. The many types of applications in a variety of social science disciplines outlined by Rummel[55] include problems of interest to chemists in management and production. Researchers in psychology and education have also benefited greatly from factor analysis. Such applications should be of value to chemists involved in education, counseling, and personnel work.

The overriding objective for writing this monograph was to equip chemists with the necessary understanding and tools to employ factor analysis in their work. That factor analysis has provided useful results even in exceptionally complicated problems suggests that applications of factor analysis in chemistry are still in an embryonic state. The range of applications discussed in the last four chapters argues well for the continued growth of factor analysis in chemistry. In a broader sense, we can anticipate that the use of diverse chemometric methods will become routine in chemistry. In particular, general procedures for attacking multidimensional problems in chemistry should result from data analyses utilizing factor analysis in conjunction with other multivariate methods such as pattern recognition, cluster analysis, and multiple regression analysis.

REFERENCES

1. M. Charton, *Chem. Tech.,* **1975,** 245.
2. B. Higman, *Applied Group-Theoretic and Matrix Methods,* Oxford University Press, Oxford, 1955.
3. E. R. Malinowski, Ph.D. thesis, Stevens Institute of Technology, Hoboken, N.J., 1961.
4. H. H. Jaffe, *Chem. Rev.,* **53,** 191 (1953).
5. P. H. Weiner, *J. Am. Chem. Soc.,* **95,** 5845 (1973).
6. J. G. Kirkwood and F. H. Westheimer, *J. Chem. Phys.,* **6,** 513 (1938).
7. S. Wold and M. Sjostrom, *Chem. Scr.,* **2,** 49 (1972).
8. J. B. Justice and T. L. Isenhour, *Anal. Chem.,* **47,** 2286 (1975).
9. R. W. Rozett and E. M. Petersen, *Anal. Chem.,* **47,** 1301 (1975).
10. R. W. Rozett and E. M. Petersen, *Anal. Chem.,* **47,** 2377 (1975).
11. R. W. Rozett and E. M. Petersen, *Anal. Chem.,* **48,** 817 (1976).
12. R. W. Rozett and E. M. Petersen, *Am. Lab.,* **9** (2), 107 (1977).
13. D. R. Burgard, S. P. Perone, and J. L. Weibers, *Anal. Chem.,* **49,** 1444 (1977).
14. G. Wernimont, *Anal. Chem.,* **39,** 554 (1967).
15. R. N. Carey, S. Wold, and J. O. Westgard, *Anal. Chem.,* **47,** 1824 (1975).
16. J. T. Edward and S. C. Wong, *J. Am. Chem. Soc.,* **99,** 4229 (1977).
17. D. G. Howery and P. Chan, submitted for publication.

18. C. L. de Ligny, N. G. van der Veen, and J. C. Van Houweilingen, *Ind. Eng. Chem. Fundam.*, **15**, 336 (1976).
19. W. R. Fawcett and T. M. Krygowski, *Can. J. Chem.*, **54**, 3283 (1976).
20. D. G. Howery, *Bull. Chem. Soc. Jap.*, **45**, 2643 (1972).
21. D. L. Duewer and H. Freiser, *Anal. Chem.*, **49**, 1940 (1977).
22. D. G. Howery and M. L. Rubinstein, submitted for publication.
23. M. A. Woodbury, R. C. Clelland, and R. J. Hickey, *Behav. Sci.*, **8**, 347 (1963).
24. C. G. Swain, H. E. Bryndza, and M. S. Swain, *J. Chem. Inf. Comput. Sci.*, **19**, 19 (1979).
25. P. H. A. Sneath, *J. Theor. Biol.*, **12**, 157 (1966).
26. G. L. Ritter and H. B. Woodruff, *Anal. Chem.*, **49**, 2116 (1977).
27. A. Cammarata and G. K. Menon, *J. Med. Chem.*, **19**, 739 (1976).
28. G. K. Menon and A. Cammarata, *J. Pharm. Sci.*, **66**, 304 (1977).
29. M. L. Weiner and P. H. Weiner, *J. Med. Chem.*, **16**, 655 (1973).
30. A. Gauthier, J. Zurli, B. C. Cros, and H. Sarles, *Rev. Eur. Etud. Clin. Biol.*, **17**, 574 (1972).
31. J. Farber, J. Tosovsky, and K. Hynck, *Act. Nerv. Super.*, **16**, 258 (1974).
32. E. R. John, *Osnovn. Probl. Elektrofiziol. Golovn. Mozga.*, **1974**, 161.
33. N. R. Bohidar, F. A. Restaino, and J. B. Schwartz, *J. Pharm. Sci.*, **64**, 966 (1975).
34. E. R. Malinowski, *Anal. Chem.*, **49**, 612 (1977).
35. D. G. Howery and J. Gelbfish, unpublished results.
36. D. J. Alpert and P. K. Hopke, Proc. Conf. Quality Assurance Environ Meas., Denver, Colo., Nov. 1978, p. 204.
37. P. K. Hopke, *J. Environ. Sci. Health,* **A11** (6), 367 (1976).
38. P. K. Hopke, E. S. Gladney, G. E. Gordon, W. H. Zoller, and A. G. Jones, *Atmos. Environ.*, **10**, 1015 (1976).
39. P. D. Gaarenstroom, S. P. Perone, and J. L. Moyers, *Environ. Sci. Technol.*, **11**, 795 (1977).
40. E. J. Knudson, D. L. Duewer, G. D. Christian, and T. V. Larson, in B. R. Kowalski, Ed., *Chemometrics: Theory and Applications,* ACS Symp. Ser. 52, American Chemical Society, Washington, D.C., 1977, p. 80.
41. I. H. Blifford and G. O. Meaker, *Atmos. Environ.*, **1**, 147 (1967).
42. J. T. Peterson, *Atmos. Environ.*, **4**, 501 (1970); *ibid.*, **6**, 433 (1972).
43. L. B. Lave and E. P. Seskin, *Air Pollution and Human Health,* Johns Hopkins University Press, Baltimore, Md., 1977, p. 33.
44. P. W. Linton, D. F. S. Natusch, P. K. Hopke, and R. L. Solomon, Proc. 4th Conf. Sensing Environ. Pollutants, New Orleans, La., Nov. 1977, p. 221.
45. P. H. Weiner, *Chem. Tech.,* **1977**, 321.
46. K. Yoshizawa, T. Ishikawa, M. Kinoshita, A. Takeda, and I. Fujie, *Nippon Jozo Kyokai Zasshi,* **69**, 581 (1974).
47. L. Reiner and A. Piendle, *Brauwissenschaft,* **27**, 1 (1974).
48. H. Martens, Y. Solberg, L. Roer, and E. Vold, *Potato Res.*, **18**, 515 (1975).
49. H. Martens and K. E. B. Knudsen, *Cereal Chem.,* in press.
50. J. Imbrie, Tech. Rep. No. 6, ONR Task No. 389-135, Northwestern University, Evanston, Ill., 1963.
51. K. M. Dawson and A. J. Sinclair, *Econ. Geol.,* **69**, 404 (1974).
52. M. E. Magar and P. W. Chuin, *Biophys. Chem.*, **1**, 18 (1973).
53. R. R. Sokal and P. H. A. Sneath, *Principles of Numerical Taxonomy,* W. H. Freeman, San Francisco, 1963, Chap. 7.
54. D. F. Toerin, *Water Res.,* **3**, 129 (1969).
55. R. J. Rummel, *Applied Factor Analysis,* Northwestern University Press, Evanston, Ill., 1970, Chap. 24.

BIBLIOGRAPHY

Adcock, C. J., *Factorial Analysis for the Non-Mathematician,* Melbourne University Press, Victoria, Australia, 1954.

Ahmavaara, Y., and T. Markkanen, *The Unified Factor Model,* Finnish Foundation for Alcoholic Studies, Helsinki, 1958.

Anderson, T. W., *An Introduction to Multivariate-Statistical Analysis,* Wiley, New York, 1958.

Burt, C. C., *The Factors of the Mind,* Macmillan, New York, 1941.

Cattell, R. B., *Factor Analysis,* Harper, New York, 1952.

Cattell, R. B., *The Scientific Use of Factor Analysis in Behavioral and Life Sciences,* Plenum Press, New York, 1978.

Child, D., *The Essentials of Factor Analysis,* Holt, Rinehart and Winston, New York, 1970.

Comrey, A. L., *A First Course in Factor Analysis,* Academic Press, New York, 1973.

Fruchter, B., *Introduction to Factor Analysis,* rev. ed., D. Van Nostrand, Princeton, N.J., 1964.

Gorsuch, R. L., *Factor Analysis,* Saunders, Philadelphia, 1974.

Guertin, W. H., and J. Bailey, *Introduction to Modern Factor Analysis,* Edwards Brothers, Ann Arbor, Mich., 1970.

Harman, H. H., *Modern Factor Analysis,* rev. 1st ed., University of Chicago Press, Chicago, 1976.

Henrysson, S., *Applicability of Factor Analysis in the Behavioral Sciences,* Almquist & Wiksell, Stockholm, 1960.

Horst, P., *Factor Analysis of Data Matrices,* Holt, Rinehart and Winston, New York, 1965.

Hotzinger, K. J., and H. H. Harman, *Factor Analysis,* University of Chicago Press, Chicago, 1941.

Joreskog, K. G., *Statistical Estimation in Factor Analysis,* Almquist & Wiksell, Stockholm, 1963.

Lawley, D. N., and A. E. Maxwell, *Factor Analysis as a Statistical Method,* 2nd ed., Butterworths, London, 1971.

Lawlis, G. F., and D. Cahtfield, *Multivariate Approaches for the Behavioral Sciences,* Texas Tech University Press, Lubbock, Tex., 1974.

Mulark, S. A., *The Foundations of Factor Analysis,* McGraw Hill, New York, 1972.

Rao, C. R., *Linear Statistical Inference and Its Applications,* Wiley, New York, 1965.

Rummel, R. J., *Applied Factor Analysis,* Northwestern University Press, Evanston, Ill., 1970.

Spearman, C., *The Abilities of Man,* Macmillan, New York, 1927.

Stephenson, W., *The Study of Behavior,* University of Chicago Press, Chicago, 1953.

Thomson, G., *The Factorial Analysis of Human Ability,* Houghton Mifflin, Boston, 1951.

Thurston, L. L., *The Vectors of the Mind,* University of Chicago Press, Chicago, 1935.

Thurston, L. L., *Multiple-Factor Analysis,* University of Chicago Press, Chicago, 1947.

Van der Geer, J. P., *Introduction to Multivariate Analysis for the Social Sciences,* W. H. Freeman, San Francisco, 1971.

Author Index

241

Subject Index

Pages in *italics* indicate definitions.